BIOLOGICAL, PSYCHOLOGICAL AND SOCIAL FOUNDATION OF SEX ROLES

BIOLOGICAL, PSYCHOLOGICAL AND SOCIAL FOUNDATION OF SEX ROLES

R Verma

MD PUBLICATIONS PVT LTD
NEW DELHI
www.mdppl.com

Published by :

MD Publications Pvt Ltd
"MD House", 11, Darya Ganj,
New Delhi - 110 002
Phone : +91-11-45355555
E-mail : contact@mdppl.com
Website: www.mdppl.com

ISBN: 978-81-7533-355-0

Publisheded and Printed by Mr. Pranav Gupta on behalf of **MD Publications Pvt Ltd** at Times Press, Delhi.

PREFACE

In our society today, men and women perform distinctly different roles which are based on nothing more than their biological gender. Although these roles do not hold true for each individual, the majority of people live out their lives in accordance with these extremely pervasive roles.

Sex roles can be defined as a set of behaviors and characteristics that are standard for each gender in a society. Sex role stereotypes are widely-held beliefs about those behaviors and characteristics. The stereotypes to a large extent become the roles. Society forces people into certain roles simply by expecting that those roles are proper and enforcing them. It has been recently and frequently proposed that the currently championed roles are limiting and damaging to all involved, men and women alike, from the time that they are children. This view holds that our sex roles are solely the product of the society in which we live, and that their incompatibility with the reality of individual personalities causes pain and stress for many people, as do other forms of oppression and stereotyping. Most of the research in this area has been based on studies which point out the inequality and subordinate position of women in Western society.

The biological view of gender roles states that the differentiated gender roles which exist in our society are the products of our evolution, and are inextricably linked with abilities predominant in one gender or the other which are determined biologically. The roles prescribed for each sex are based on physical abilities and properties of that sex, such as intelligence, brain lateralization, and differing hormone levels. This view was the accepted one throughout history, and has only recently been challenged. Implicit within this view is the belief that these roles are both desirous and immutable. Ways in which this view could be substantiated include looking at gender role similarities between animals and humans, and examining sex differences among infants (Singleton, 1987). If substantial sex role similarities are found between animals and humans, or if infants are in fact found to exhibit traits traditionally assigned to their gender, this could point toward the fact that such abilities and roles are biologically determined.

Contrary to the biological determinism perspective, the socio-cultural view of gender roles states that gender roles and stereotypes develop within a culture and are then perpetuated by that culture. Individuals within the culture are expected to conform to these norms, and are socialized in manners which constantly reinforce the beliefs and behavior which are prescribed and presupposed for them.

Current and past changes in sex roles can moreover be explained by the fact that, since cultures change, what roles are adaptive to each culture will also change over time, and should do so. The socio-cultural view, influenced to a great extent by feminism, further states that the current female sex roles in our society are psychologically damaging to women, in that they promote as desirable behaviors and beliefs which are incompatible with reality and are maladaptive to psychological adjustment. The male gender roles are also damaging to men for the same reasons.

I am very grateful to Mr. Pranav Gupta, Director of MD Publications Pvt Ltd, New Delhi, for his kind cooperation and support for this book.

R Verma

CONTENTS

1

Gender Role

A gender role is defined as a set of perceived behavioral norms associated particularly with males or females, in a given social group or system. It can be a form of division of labour by gender. It is a focus of analysis in the social sciences and humanities. Gender is one component of the gender/sex system, which refers to "The set of arrangements by which a society transforms biological sexuality into products of human activity, and in which these transformed needs are satisfied" (Reiter 1975: 159). Almost all societies, to a certain effect, have a gender/sex system, although the components and workings of this system vary markedly from society to society.

Gender refers to an individual's "psychological type". It is acquired through experience. An individual can be viewed as either masculine or feminine. Gender role refers to the attitudes and behaviors that class a person's stereotypical identity, e.g. women cook and clean, men fix cars. To associate one as either masculine or feminine is identifying with gender.

Researchers recognize that the concrete behavior of individuals is a consequence of both socially enforced rules and values, and individual disposition, whether genetic, unconscious, or conscious. Some researchers emphasize the objective social system and others emphasize subjective orientations and dispositions. Creativity may cause the rules and values to change over time. Cultures and societies are dynamic and ever changing, but there has been extensive debate as to how, and how fast, they may change. Such debates are especially contentious when they involve the gender/sex system, as people have widely differing views about how much gender depends on biological sex.

Where do Gender Roles Come From

A person's sexuality comes from within him or her, making a person heterosexual, homosexual, bisexual, or asexual, depending on the partners he or she is (or is not) attracted to. Unlike sexuality, however, gender roles are imposed from without, through a variety of social influences. Formed during the socialization phases of childhood and adolescence, gender role issues influence people throughout their lives; conflict can arise when some one does not feel at ease with his or her gender role.

The first and one of the strongest influences on a person's perceived gender role is his or her parents. Parents are our first teachers- not only of such basic skills as talking and walking, but also of attitudes and behavior. Some parents still hold traditional definitions of maleness and femaleness and what kind of activities is appropriate for each.

Parents start early in treating their baby boys and baby girls differently. Although baby boys are more likely to die in infancy than girls, and are actually more fragile as infants than girls are, studies have shown that parents tend to respond more quickly to an infant daughter's cries than they are to those of an infant son. Parents also tend to cuddle girls more than they do boys. They are also more likely to allow boys to try new things and activities-such as learning to walk and explore-than they are girls; parents tend to fear more for the safety of girls.

According to Dr. Benjamin Spock, people are likely to appreciate girls' cuteness and boys' achievements. For example, a girl may receive the comment, "You look so pretty!" for the outfit she is wearing. While this compliment isn't harmful in itself, repeated over and over the message the girl gets is that she is most appreciated for her looks, not for what she can do. Boys, on the other hand, are praised for what they can do-"Aren't you a big boy, standing up by yourself!" Many parents encourage and expect boys to be more active, to be more rough-and-tumble in their play than girls. A boy who does not like rough play (and so goes against the gender role he has been assigned) may be labeled a "sissy." A girl who prefers active play to more passive pursuits may be called a "tomboy."

Children look to their parents for examples and role models. If a girl sees her mother taking part in physical activities, for example, she will grow up with the idea that it's okay for girls to play sports. If a boy sees his father helping to take care of the new baby, he will integrate this image of "daddy as care giver" into his developing definition of masculinity.

But just as parents can provide positive role models, so too can they serve as negative role models. For example, children who grow up with parents who are in an abusive relationship have been found to repeat the same pattern as adults: male children of abusive husbands often grow up to abuse their own wives, and daughters of abused wives can grow up to be victims of domestic violence, because their parents have shown them that this is "normal."

Children develop their gender identity (knowing whether they are male or female) by the age of three. As preschoolers, they use some sexual stereotypes to help them differentiate between men and women-for example, to a preschooler, long hair may mean "female" and short hair, "male."

Another influence and reinforcement of gender roles comes from the toys children play with. During their infancy and toddler hood, children get most of their toys from parents and other family members; their choice of toys supports their own view of gender roles. For example, parents may give their little girl a doll to sleep with, while the boy gets a teddy bear. A grandparent may give a grandson a toy truck but never consider giving the same to a granddaughter. Such gifts set children up early on for the roles they are expected to play.

As they get older, children are influenced in their choice of toys by television. Remote-controlled vehicles, although they can be equally enjoyed by males or females, are generally targeted at boys by advertisers. Girls are the advertising targets of the manufacturers of dolls, craft kits, and so on; advertisers are careful not to call boys' toys "dolls"-they're "action figures"! Again and again, we see toys and toy advertisement reinforcing the traditional gender roles: boys are active and adventurous, while girls are passive and mothering. Parents need to be aware of the messages TV advertisements and toys present to their children. They need to help them understand and reconcile the person they are with the sexual stereotypes they may see on TV and in other media.

Nevertheless, parents can and do reinforce sexual stereotypes, whether deliberately or unwittingly. Not wanting to see a daughter fall and get hurt, a mother may forbid her from climbing trees-although her brother is allowed to do so with gleeful abandon, and his bumps and bruises are taken in stride. Clothing manufacturers produce (and parents buy) clothing in gender-neutral shades such as yellow and green, but the traditional blue for boys and pink for girls are still favorites. Even the cultural habit of assigning pink to girls and blue to boys raises a question-what's to become of the boy who genuinely likes the color pink? This question leads us to another group that has strong influence over gender roles: peers.

Peer pressure is a means of reinforcing a culture's traditional gender roles. It can come in the form of taunting or teasing a child who does not fit the traditional gender roles that other children in the peer group have been exposed to, even to the point of excluding that child from group activities.

Peers react more positively to children who fit traditional gender roles. For example, the *Washington Post* reported the case of a five-year-old boy whose favorite color was pink, and as a result, when the time came to buy him his first bicycle, he naturally wanted it to be pink. The parents had no problem with this, and the boy even told the salesman (who tried to tell him that boys should ride blue or red bikes) that color was just color. The ones who teased him about his bike were not the other boys, but the girls in the neighborhood. Not long after, the boy stopped telling other people that pink was his favourite color.

Resisting such teasing takes a strong ego, something that takes many people years to develop. In a study conducted at Suffolk University in Boston, researcher Krisanne Bursik studied the ego development of 209 undergraduates and compared the results to gender-related traits. She found that students who were more likely to express non-traditional gender role traits had higher levels of ego development. She found that among male students, those who had less-developed egos viewed high levels of traditional masculinity as the ideal. She noted that in these men, "gender role conflict may occur for men when rigid, sexist or restrictive gender roles, learned during socialization, result in personal restriction, devaluation, or violation of others or self." However, Bursik's research was unable to answer the chicken-or-the egg question: which comes first? Do people who have strong, well-developed egos feel free to go against traditional gender roles? Or does early exposure to alternatives where gender roles are concerned lead people to develop strong egos?

Gender roles are also reinforced by school. Teachers and school administrators have great influence as they pass along cultural information and expectations. In school, children are expected to sit still, read, and be quiet. Such expectations may have been part of the gender role that a child has been learning from the parents, especially if the child is a girl. But for a boy who has been encouraged to be loud and boisterous prior to starting school, these expectations can lead to trouble. In fact, some researchers maintain that all boys face difficulty with expectations such as these because the structure of their brains makes them less able to meet these expectations than girls are.

Although viewed with dismay by some, schools around the country are trying out single-sex classes. About three percent of the children in the Baltimore school system are enrolled in such classes. Not only have the children's scores on performance assessment tests increased, but the single-sex classes have given advantages to the children. Boys who had been shy to speak in class do well in them, and girls develop leadership skills. Students at other single-sex schools have been found to be less susceptible to social pressure. Are these successes due to the positive gender role expectations their teachers present, or does going to a single-sex school or class eliminate part of peer pressure? The research is continuing.

Talcott Parson's view of Gender Roles

Working in the United States, Talcott Parsons developed a model of the nuclear family in 1955. (At that place and time, the nuclear family was the prevalent family structure.) It compared a strictly traditional view of gender roles (from an industrial-age American perspective) to a more liberal view.

The Parsons model was used to contrast and illustrate extreme positions on gender roles. Model A describes total separation of male and female roles, while Model B describes the complete dissolution of barriers between gender roles. (The examples are based on the context of the culture and infrastructure of the United States.)

Model A - Total Role Segregation

Education-Gender-specific education; high professional qualification is important only for the man-Co-educative schools, same content of classes for girls and boys, same qualification for men and women.

Profession-The workplace is not the primary area of women; career and professional advancement is deemed unimportant for women. For women, career is just as important as for men; therefore equal professional opportunities for men and women are necessary.

Housework-Housekeeping and child care are the primary functions of the woman; participation of the man in these functions is only partially wanted.-All housework is done by both parties to the marriage in equal shares.

Decision making-In case of conflict, man has the last say, for example in choosing the place to live, choice of school for children, buying decisions-Neither partner dominates; solutions do not always follow the principle of finding a concerted decision; status quo is maintained if disagreement occurs.

Child care and education-Woman takes care of the largest part of these functions; she educates children and cares for them in every way-Man and woman share these functions equally.

However, these extreme positions are rarely found in reality; actual behavior of individuals is usually somewhere between these poles. The most common 'model' followed in real life in the United States and Britain is the 'model of double burden.' According to the interactionist approach, roles (including gender roles) are not fixed, but are constantly negotiated between individuals. In North America and southern South America, this is the most common approach among families whose business is agriculture.

Gender roles can influence all kinds of behavior, such as choice of clothing, choice of work and personal relationships; e.g., parental status

Gender Roles and Socialization

The process through which the individual learns and accepts roles is called socialization. Socialization works by encouraging wanted and discouraging unwanted behavior. These sanctions by agents of socialization such as the family, schools, and the media make it clear to the child what is expected of the child by society. Mostly, accepted behavior is not produced by outright reforming coercion from an accepted social system. In some other cases, various forms of coercion have been used to acquire a desired response or function.

In the majority of traditional and developmental social systems, an individual has a choice as to what he or she should or should not conform to as a representative of the socialization process. Through this process an individual willingly accepts the consequences whether they are beneficial or negative, minor or severe depending on the way an individual copes, and the extent to which he or she must modify his or her natural behavior to conform to society's accepted behavioral standards. Typical encouragements and expectations of gender role behavior are not as powerful a difference and reforming social trait as they were a century ago. Such developments and traditional refinements are still a socialization process to and within family values, peer pressures, at the employment centers and in every social system communication medium.

Still, once someone has accepted certain gender roles and gender differences as expected socialized behavioral norm, his or her behavior traits become part of his or her perceived responsibilities. Influential roles in gender relationships on a personal and social level to the individual's own socializing role or self-concept. Sanctions on unwanted behavior and role conflict can be stressful.

GENDER ROLES WITH HOMOGENIZATION VS. ETHNO CONVERGENCE DIFFERENCE

It is claimed that even in monolingual, industrial societies like urban North America, some individuals do cling to a "modernized" primordial identity, apart from others and with this a more diverse gender role is recognized or developed. Some intellectuals, such as Michael Ignatieff, argue that convergence of a general culture does not directly entail a similar convergence in ethnic, social and self identities. This can become evident in social situations, where people divide into separate groups by gender roles and cultural alignments, despite being of an identical *"super-ethnicity"*, such as nationality.

Within each smaller ethnicity, individuals may tend to see it perfectly justified assimilating with other cultures including sexuality and some others view assimilation as wrong and incorrect for their culture or institution. This common theme, representing dualist opinions of ethno convergence itself, within single ethnic or common values groups is often manifested in issues of sexual partners and matrimony, employment preferences, etc. These varied opinions of ethno convergence represent themselves in a spectrum; assimilation, homogenization, acculturation, gender identities and cultural compromise are commonly used terms for ethno convergence which flavor the issues to a bias.

Often it is in a secular, multi-ethnic environment that cultural concerns are both minimalized and exacerbated; Ethnic prides are boasted, hierarchy is created ("center" culture versus "periphery") but on the other hand, they will still share a common "culture", and common language and behaviors. Often the elderly, more conservative-in-association of a clan, tend to reject cross-cultural associations, and participate in ethnically similar community-oriented activities. Xenophobes tend to think of cross-cultural contact as a component of gender and assimilation and see this as terrible.

The idea that differences in gender roles originate in differences in biology has found support in parts of the scientific community. 19th-century anthropology sometimes used descriptions of the imagined life of Paleolithic hunter-gatherer societies for evolutionary explanations for gender differences. For example, those accounts maintain that the need to take care of offspring may have limited the females' freedom to hunt and assume positions of power.

More recently, sociobiology and evolutionary psychology have explained those differences in social roles by treating them as adaptations. This approach, too, is considered controversial.

Due to the influence of (among others) Simone de Beauvoir's feminist works and Michel Foucault's reflections on sexuality, the idea that gender was unrelated to sex gained ground during the 1980s, especially in sociology and cultural anthropology. This view claims that a person could therefore be born with male genitals but still be of feminine gender. In 1987, R.W. Connell did extensive research on whether there are any connections between biology and gender role and concluded that there were none. Most scientists reject Connell's research because concrete evidence exists proving the effect of hormones on behavior. However, hormone levels vary, and disorders can cause an intersex status. The debate continues to rage on. Simon Baron-Cohen, a Cambridge Univ. professor of psychology and psychiatry, claims that "the female brain is predominantly hard-wired for empathy, while the male brain is predominantly hard-wired for understanding and building systems."

Dr. Sandra Lipsitz Bem is a psychologist who developed the gender schema theory to explain how individuals come to use gender as an organizing category in all aspects of their life. It is based on the combination of aspects of the social learning theory and the cognitive-development theory of sex role acquisition. In 1971, she created the Bem Sex Role Inventory to measure how well you fit into your traditional gender role by characterizing your personality as masculine, feminine, androgynous, or undifferentiated. She believed that through gender-schematic processing, a person spontaneously sorts attributes and behaviors into masculine and feminine categories. Therefore, an individual processes information and regulate their behavior based on whatever definitions of femininity and masculinity their culture provides.

The current trend in Western societies toward men and women sharing similar occupations, responsibilities and jobs suggests that the sex one is born with does not directly determine one's abilities. While there are differences in average capabilities of various kinds (e.g., physical strength) between the sexes, the capabilities of some members of one sex will fall within the range of capabilities needed for tasks conventionally assigned to the other sex.

In addition, research at the Yerkes National Primate Research Center has also shown that gender roles may be biological among primates. Yerkes researchers studied the interactions of 11 male and 23 female Rhesus monkeys with human toys, both wheeled and plush. The males played mostly with the wheeled toys while the females played with both types equally. Psychologist Kim Wallen has, however, warned against over interpeting the results as the color and size of the toys may also be factors in the monkey's behavior.

CHANGING ROLES

A person's gender role is composed of several elements and can be expressed through clothing, behaviour, choice of work, personal relationships and other factors. These elements are not concrete and have evolved through time (for example women's trousers).

Traditionally only feminine and masculine gender roles existed, however, over time many different acceptable male or female gender roles have emerged. An individual can either identify themselves with a subculture or social group which results in them having diverse gender roles. Historically, for example, eunuchs had a different gender role because their biology was changed. A woman publicly witnessing at a Quaker meeting seemed an extraordinary feature of the Religious Society of Friends, worth recording for a wider public.

ANDROGYNY

Androgyny, a term denoting the display of both male and female behaviour, also exists. Many terms have been developed to portray sets of behaviors arising in this context. The masculine gender role in the West has become more malleable since the 1950s. One example is the "sensitive new age guy", which could be described as a traditional male gender role with a more typically "female" empathy and associated emotional responses. Another is the metro sexual, a male who adopts or claims to be born with similarly "female" grooming habits. Some have argued that such new roles are merely rebelling against tradition more so than forming a distinct role. However, traditions regarding male and female appearance have never been concrete, and men in other eras have been equally interested with their appearance.

The popular conceptualization of homosexual men, which has become more accepted in recent decades, has traditionally been more androgynous or effeminate, though in actuality homosexual men can also be masculine and even exhibit machismo characteristics. One could argue that since many homosexual men and women fall into one gender role or another or are androgynous, that gender roles are not strictly determined by a person's physical sex. Whether or not this phenomenon is due to social or biological reasons is debated. Many homosexual people find the traditional gender roles to be very restrictive, especially during childhood. Also, the phenomenon of intersex people, which has become more publicly accepted, has caused much debate on the subject of gender roles. Many intersexual people identify with the opposite sex, while others are more androgynous. Some see this as a threat to traditional gender roles, while others see it as a sign that these roles are a social construct, and that a change in gender roles will be liberating.

According to sociology research, traditional feminine gender roles have become less relevant and hollower in Western societies since industrialization started. For example, the cliché that women do not follow a career is obsolete in many Western societies. On the other hand, in the media there are attempts to portray women who adopt an extremely classical role as a subculture. Women take on many roles that were traditionally reserved for men, as well as behaviors and fashions, which may cause pressure on many men to be more masculine and thus confined within an even smaller gender role, while other men react against this pressure. For example, men's fashions have become more restrictive than in other eras, while women's fashions have become broader. One consequence of social unrest during the Vietnam War era was that men began to let their hair grow to a length that had previously (within recent history) been considered appropriate only for women. Somewhat earlier, women had begun to cut their hair to lengths previously considered appropriate only to men.

CULTURE AND GENDER ROLES

During World War II, women filled job positions some of which would otherwise be male dominated.

Ideas of appropriate behavior according to gender vary among cultures and era, although some aspects receive more widespread attention than others. An interesting case is described by R.W. Connell in *Men, Masculinities and Feminism:*

"There are cultures where it has been normal, not exceptional, for men to have homosexual relations. There have been periods in 'Western' history when the modern convention that men suppress displays of emotion did not apply at all, when men were demonstrative about their feeling for their friends. Mateship in the Australian outback last century is a case in point."

Other aspects, however, may differ markedly with time and place. In pre-industrial Europe, for example, the practice of medicine (Other than midwifery) was generally seen as a male prerogative. However, in Russia, health care was more often seen as a feminine role. The results of these views can still be seen in modern society, where European medicine is most often practiced by men, while the majority of Russian doctors are women.

In many other cases, the elements of convention or tradition seem to play a dominant role in deciding which occupations fit in with which gender roles. In the United States, physicians have traditionally been men, and the few people who defied that expectation received a special job description: "woman doctor". Similarly, there were special terms like "male nurse", "woman lawyer", "lady barber", "male secretary," etc. But in the former Soviet Union countries, medical doctors are predominantly women, and in Germany and Taiwan it is very common for all of the barbers in a barber shop to be women. Also, throughout history, some jobs that have been typically male or female have switched genders. For example, clerical jobs used to be considered men's jobs, but when several women began filling men's job positions due to World War II, clerical jobs quickly became dominated by women. It became more feminized, and women workers became known as "typewriters" or "secretaries". There are many other jobs that have switched gender roles. Many jobs are continually evolving as far as being dominated by women or men.

In Western society, people whose gender appears masculine are sometimes ridiculed for exhibiting what the society regards as a woman's gender role. For instance, someone with a masculine voice, a five o'clock shadow (or a fuller beard), an Adam's apple, etc., wearing a woman's dress and high heels, carrying a purse, etc., would most likely draw ridicule or other unfriendly attention in ordinary social contexts (the stage and screen excepted). It is seen by some in that society that such a gender role for a man is not acceptable. This, and other societies, imposes expectations on the behaviour of the members of society, and specifically on the gender roles of individuals, resulting in prescriptions regarding gender roles.

TRANSGENDERED AND INTERSEXED PEOPLE

As long as a person's perceived physiological sex is consistent with that person's gender identity, the gender role of a person is so much a matter of course in a stable society that people rarely even think of it. Only in cases where, for whatever reason, an individual has a gender role that is inconsistent with his or her sex will the matter draw attention.

There are cases wherein the external genitalia of a person, that person's perceived gender identity, and/or that person's gender roles are not consistent. People naturally, but too easily, assume that if a person has a penis, scrotum, etc., then that person is chromosomally male (i.e., that person has one X chromosome and one Y chromosome), and that the person, in introspection, feels like a male.

In one example, a person may have a penis and scrotum, but may be a female (With XX chromosomal sexual identity and with normal female sexual organs internally). When that person reaches puberty, "his" breasts may enlarge to ordinary female proportions, and "he" may begin to menstruate, passing menstrual blood through "his" penis. In addition, this person may have always accepted a gender identity that is consistent with "his" external genitalia or with "her" internal genitalia. When the true sex of the individual becomes revealed at puberty, the individual and/or the community will be forced to reconsider what gender role is to be considered appropriate. Biological conditions that cause a person's physiological sex to be not easily determined are collectively known as intersex.

Another example to consider is transgender people, who mix gender roles to form a personally comfortable androgynous combination or transcend the scheme of gender roles completely, regardless of their physiological sex. Transgender people can also be physically androgynous or identify as androgynous. Transsexualism also exists, where a person who is born as one sex and is brought up in that sex, but has gender identity of the opposite sex and wishes to live as that sex (but does not necessarily wish to adopt that particular gender role, depending upon whether the person in question believes that gender roles are innate).

When we consider these more unusual products of nature's inventiveness, the simple picture that we saw originally, in which there was a high degree of consistency among external genitalia, gender identity, and gender role, then dissolves into a kind of jigsaw puzzle that is difficult to put together correctly. The extra parts of this jigsaw puzzle fall into two closely related categories, atypical gender identities and atypical gender roles.

In Western society, there is a growing acceptance of intersexed and transgendered people. However, there are some who do not accept these people and may react violently and persecute them: this kind of negative value judgment is sometimes known as transphobia.

Nevertheless, such cases of mismatch between a person's physiology, identity and role are rare. A large majority of people have matching genitalia and gender identities. For many people their gender role is commensurate with their genitalia.

GENDER ROLES AND FEMINISM

For approximately the last 100 years women have been fighting for the same rights as men (especially around the turn from 19th to 20th century with the struggle for women's suffrage and in the 1960s with second-wave feminism and radical feminism) and were able to make changes to the traditionally accepted feminine gender role. However, most feminists today say there is still work to be done.

Numerous studies and statistics show that even though the situation for women has improved during the last century, discrimination is still widespread: Women earn a smaller percentage of aggregate income than men, occupy lower-ranking job positions than men and do most of the housekeeping work. However, there may be some reason for this, as some studies have indicated that many jobs which were traditionally perceived to be male-dominated usually have longer hours, necessitate long periods of exposure to the elements, are higher risk and require a fair amount of physical strength.

A recent article in *The New York Times* indicated that gender roles are still prevalent in many upscale restaurants. A restaurant's decor and menu typically play into which gender frequents which restaurant. Whereas Cru, a restaurant in New York's, Greenwich Village, "decorated in clubby brown tones and distinguished by a wine list that lets high rollers rack up breathtaking bills," attracts more men than women, places like Mario Batali's, Otto, serves more women than men, as a result that the restaurant has been "designed to be more approachable, with less swagger." Servers of both men and women at the same table still often go with the assumption that the male is the go-to person, as far as who receives the check and makes the wine decisions, but this appears to be a trend that is being used with more caution, especially with groups of younger people. Restaurants that used to cater to more men or women are now also trying to change their decor in the hopes of attracting broader equity.

SEXUAL ORIENTATION AND GENDER ROLES

Traditional gender roles include male attraction to females, and vice versa. Gay, lesbian and bisexual people, among others, usually don't conform to these expectations. An active conflict over the cultural acceptability of non-heterosexuality rages worldwide. The belief or assumption that heterosexual relationships and acts are "normal" is described - largely by the opponents of this viewpoint - as heterosexism or in queer theory, heteronormativity.

Perhaps it is an attempt to reconcile this conflict that leads to a common assumption that one same-sex partner assumes a pseudo-male gender role and the other assumes a pseudo-female role. For a gay male relationship, this might lead to the assumption that the "wife" handled domestic chores, was the receptive sexual partner in anal sex, adopted effeminate mannerisms, and perhaps even dressed in women's clothing.

This assumption is flawed, as many homosexual couples tend to have more equal roles, and the effeminate behavior of some gay men is usually not adopted consciously, and is often more subtle. Feminine or masculine behaviors in some homosexual people might be a product of the socialization process, adopted unconsciously due to stronger identification with the opposite sex during development. The role of both this process and the role of biology are debated.

The existence of these separate identities (dominant masculine vs more passive feminine), where present, can establish the dynamics of the relationship, according to the heterosexual patterns; this is not always the case, especially in relationships with less clearly defined sexual/identity roles. A related assumption is that all androphilic people, including gay men, should or do adopt feminine mannerisms and other gender-role elements, and that all gynophilic people, including lesbians, should or do adopt masculine mannerisms and other gender-role elements; it is unclear how bisexuality fits into this framework, but it can be assumed they have a tendency towards both gender roles as they do in sexuality, towards both sexes. However, this idea is based on generalizations of homosexual people, which tend to be biased, as feminine gays and masculine lesbians are more widely visible than masculine gays or feminine lesbians.

Same-sex domestic partners also challenge traditional gender roles because it is impossible to divide up household responsibilities if both partners attempt to fill the same gender role. Like all live-in couples, same-sex partners usually do come to some arrangement with regard to household responsibilities. Sometimes these arrangements do assign traditional female responsibilities to one partner and traditional male responsibilities to the other, but non-traditional divisions of labor are also quite common. For instance, cleaning and cooking, traditionally both female responsibilities might be assigned to different people. Some people do adopt the sexual role of bottom or top, due to thei. own sexual identity or for convenience; but this is not universal, and does not necessarily correspond to assignment of household responsibilities.

Cross-dressing is also quite common in gay and lesbian culture, but it is usually restricted to festive occasions, though there are people of all sexual orientations who routinely engage in various types of cross-dressing, either as a fashion statement or for entertainment. Distinctive styles of dress, however, are commonly seen in gay and lesbian circles. These fashions sometimes emulate the traditional styles of the opposite gender (For example, lesbians who wear t-shirts and boots instead of skirts and dresses, or gay men who wear clothing with traditionally feminine elements, including displays of jewelry or coloration), but others do not. Fashion choices also do not necessarily align with other elements of gender identity. Some fashion and behavioral elements in gay and lesbian culture are novel, and do not really correspond to any traditional gender roles. For example, the popularity of rainbow jewelry, or the gay techno/dance music subculture. In addition to the stereotypically effeminate one, another significant gay male subculture is homomasculinity, emphasizing certain traditionally masculine or hyper masculine traits. (See Sexuality and gender identity-based cultures.) These may be natural traits for an individual, or they may be adopted to conform to mainstream culture or to distance one from the more effeminate gays, as masculinity is often seen as a positive trait among gay men. A quick look through gay personals websites, such as Gaydar, will reveal that many gay men's personals are discouraging effeminate men from replying to their ad. (Fat, effeminate or bald, please do not disturb).

The term *dyke*, commonly used to mean *lesbian*, sometimes carries associations of a butch or masculine identity, and the variant *bulldyke* certainly does. Other gender-role-charged lesbian terms include lipstick lesbian, chapstick lesbian, and Stone Femme. "Butch," "femme," and novel elements are also seen in various lesbian subcultures.

External social pressures may lead some people to adopt a persona which is perceived as more appropriate for a heterosexual (For instance, in an intolerant work environment) or homosexual (for instance, in a same-sex dating environment), while maintaining a somewhat different identity in other, more private circumstances. The acceptance of new gender roles in Western societies, however, is rising. However, during childhood and adolescence, gender identities which differ from the norm are often the cause of ridicule and ostracism, which often results in psychological problems. Some are able to disguise their differences, but others are not. Even though much of society has become more tolerant, gender roles are still very prevalent in the emotionally charged world of children and teenagers, which makes life very difficult for those who differ from the established norms.

THE ROLE OF IDEOLOGY IN THE ENCULTURATION OF GENDER ROLES

High levels of agreement on the characteristics different cultures to males and females reflects consensus in gender role ideology. The Netherlands, Germany and Finland are among the most egalitarian modern societies concerning gender roles, whereas the most traditional roles are found in Nigeria, Pakistan, and India. Men and women cross-culturally rate the ideal self as more masculine than their self. The American difference on spatial reasoning between males and females does not apply in all cultures. One example of where it does not is the Inuit culture in Canada. Male superiority is found in tight, sedentary, agriculturally based cultures, while females are superior in cultures that are loose, nomadic, and hunter-gathering based. US females are more conforming to others than are males. Males are more aggressive in all cultures for which data exists. This is related to, but not solely determined by, age and hormones though some researchers would suggest that women are not necessarily less aggressive than men but tend to show their aggression in more subtle and less overt ways (Bjorkqvist et al. 1994, Hines and Saudino 2003). Male aggression may be a "gender marking" issue breaking away from the instruction of the mother during adolescence. Native American gender roles depend on the cultural history of the tribe.

Gender Roles in Prison

Gender roles in male prisons go further than the "Don't drop the soap"-joke. The truth is that some prisoners, either by choice or by force, take on strict 'female roles' according to prison set guidelines. For instance, a 'female' in prison is seen as timid, submissive, passive, and a means of sexual pleasure. When entering the prison environment some inmates "turn out" on their own free will, meaning they actively pursue the 'female role' in prison to gain some form of social power and/or prestige. Other, unlucky inmates are forced to partake in 'female role' activities through coercion; the most common means being physical abuse. The inmates that are forced to "turn out" are commonly referred to as "punks". Other terms used to describe 'female' inmates are "girls", "kids", and "gumps". Some of the labels may be used as a means of describing one's ascribed status. For example, a "kid" is one that is usually dominated by their owner, or "daddy". The "daddy" is usually one with a high social status and prestige within the prison (E.g. gang leader). The "female" gender role is constructed through the mirror image of what the in-mates perceive as a male. For instance, inmates view men as having strength, power, prestige, and an unyielding personality. However, the inmates don't refer to the female guards, who have power and prestige over the inmates, as males. The female guards are commonly referred to as "dykes", "ditch lickers", and lesbians. These roles are also assumed in female prisons.

Women who enter prison society often voluntarily enter into lesbianism, as a means of protection from gangs or stronger females. In doing so, they will take on the submissive role to a dominant female in exchange for that dominating female keeping them safe. Those who do not enter voluntarily into lesbianism might at one time or another is group raped, to introduce them into that circle, and sometimes they will be referred to as sheep, meaning anyone can have them. It is to avoid that status that most female inmates choose a mate, or allow themselves to be chosen as a mate, which can make them available to only a minimal number of partners during their incarceration, as opposed to a large number. So, in a sense, an inmate undergoes a "female role" in the prison system either by choice or by yielding to excessive coercion, and it is that yielding those terms the once male inmates as "females", and which identifies the stronger females in a female prison system as "males".

Gender roles are "socially and culturally defined prescriptions and beliefs about the behavior and emotions of men and women" (Anselmi and Law 1998, p. 195). Many theorists believe that perceived gender roles form the bases for the development of gender identity. Prominent psychological theories of gender role and gender identity development include evolutionary theory (Buss 1995; Shields 1975), object-relations theory (Chodorow 1989), gender schema theory (Bem 1981, 1993) and social role theory (Eagly 1987).

Evolutionary theories of gender development are grounded in genetic bases for differences between men and women. Functionalists (e.g., Shields 1975) propose that men and women have evolved differently to fulfill their different and complementary functions, which are necessary for survival. Similarly, sociobiologists (e.g., Buss 1995) suggest that behavioral differences between men and women stem from different sexual and reproductive strategies that have evolved to ensure that men and women are able to efficiently reproduce and effectively pass on their genes. These evolutionary-based theories share similarities with the essentialist and maximalist perspectives discussed previously.

In contrast, object-relations theorists focus on the effects of socialization on gender development. For example, Nancy Chodorow (1989) emphasizes the role of women as primary caregivers in the development of sex differences. Chodorow asserts that the early bond between mother and child affects boys and girls differently. Whereas boys must separate from their mothers to form their identities as males, girls do not have to endure this separation to define their identities as females. Chodorow (1989) explains that the devalued role of women is a product of the painful process men undergoes to separate them from the female role.

Gender schema theory (Bem 1981) focuses on the role of cognitive organization in addition to socialization. This theory postulates that children learn how their cultures and/or societies define the roles of men and women and then internalize this knowledge as a gender schema, or unchallenged core belief. The gender schema is then used to organize subsequent experiences

(Bem 1993). Children's perceptions of men and women are thus an interaction between their gender schemas and their experiences. Eventually, children will incorporate their own self-concepts into their gender schema and will assume the traits and behaviors that they deem suitable for their gender.

Alice Eagly (1987) offers yet another explanation of gender development that is based on socialization. Eagly's social role theory suggests that the sexual division of labor and societal expectations based on stereotypes produce gender roles. Eagly (1987) distinguishes between the communal and agentic dimensions of gender-stereotyped characteristics. The communal role is characterized by attributes, such as nurturance and emotional expressiveness, commonly associated with domestic activities, and thus, with women. The agentic role is characterized by attributes such as assertiveness and independence, commonly associated with public activities, and thus, with men. Behavior is strongly influenced by gender roles when cultures endorse gender stereotypes and form firm expectations based on those stereotypes (Eagly 1987).

As Eagly suggests, gender roles are closely linked with gender stereotypes. Stereotypes are "overgeneralized beliefs about people based on their membership in one of many social categories" (Anselmi and Law 1998, p. 195). Gender stereotypes vary on four dimensions: traits, role behaviors, physical characteristics, and occupations (Deaux and Lewis 1983). For example, whereas men are more likely to be perceived as aggressive and competitive, women are more likely to be viewed as passive and cooperative. Traditionally, men have been viewed as financial providers, whereas women have been viewed as caretakers. Physical characteristics and occupations have also been considered consistent or inconsistent with masculine or feminine roles.

Traditional gender stereotypes are most representative of the dominant (white, middle-class) culture. Hope Landrine (1999) asserts that although race and social class may not be mentioned when inquiring about gender stereotypes, most people will make assumptions about these categories. Her research suggests that when race and social classes are specified, different gender stereotypes emerge.

Gender roles and stereotypes affect couple and family interaction. Often, for example, the division of household labor is based on gender. Traditionally, white women in heterosexual couples remained at home and completed most of the domestic labor, while their male partners worked outside the home to provide the family income. Although women have increasingly joined the workforce over the past thirty years, they continue to do the majority of the household labor. Lawrence Kurdek (1993) studied white, heterosexual, gay, and lesbian couples without children. He found that heterosexual and gay couples were more likely than lesbian couples to divide household labor so that one partner did the majority of the work. Lesbian couples were most likely to share domestic tasks or take turns doing the tasks (Kurdek 1993).

Gender roles often become more differentiated when men and women become parents. Overall, women provide more direct care for and spend more time with children (Walzer 2001). This care includes taking responsibility for the mental work of gathering and processing information about infant care, delegating the tasks related to infant care, and worrying about infant health and well-being. In sum, the unequal division of both household labor and childcare, with women doing the bulk of the work, is thought to contribute to the reported lower marital satisfaction for women (Walzer 2001).

Gender roles and stereotypes affect men and women in other ways. Specifically, men and women may be judged by how well they conform to traditional stereotypes. In his theory of masculine gender role strain, Joespech Pleck (1976) asserted that boys and men are pressured to fulfill a standard of masculinity. Boys and men, for example, who do not fulfill the standard often, suffer from low self-worth (Pleck; Sonnenstein; and Ku1993). Other lifelong consequences befall men who experience traumatic socialization practices such as rites of passage that entail violence. Even men who successfully fulfill the standard of masculinity suffer psychologically or emotionally from rigid constraints on acceptable parenting roles for men (Pleck; Sonnonstein; and Ku 1993). Richard Lazur and Richard Majors (1995) contend that gender role strain is pronounced with men of color. Men of color must balance the dominant standards of masculinity with their cultures' standards of masculinity in an effort to fulfill both satisfactorily. In addition, men of color must overcome prejudice and other obstacles to fulfill the standards of masculinity. The result is increased gender role strain for men of color (Lazur and Majors 1995). Likewise, white women and women of color may be constrained by standards of femininity, such as the pressure tc have children.

Gender stereotypes can also affect men's and women's performance. Stereotype threat is defined as "an individual's awareness that he or she may be judged by or may self-fulfill negative stereo-types about her or his gender or ethnic group" (Lips 2001, p. 33). Research indicates that stereo-type threat can negatively affect performance by increasing anxiety. For example, Steven Spencer, Claude Steele, and Diane Quinn (1999) found that women performed significantly worse than men on a math test when the participants were led to believe that the test would probably produce gender differences. In contrast, women and men performed equally well when the participants were led to believe that the test did not produce gender differences. These findings suggest that negative stereotypes can and do negatively affect performance even when the stereotype has not been internalized or incorporated into the view of the self.

GENDER DIFFERENCES

It refers to the behaviors, attitudes, values, beliefs and so on that is considered appropriate by a particular cultural group for the males and females on the basis of their biological sex. It is also known as sex role though there is a difference because while sex role refers to whether someone lives a male or female role, gender role is how he, or she, lives that role. As a result of socio-cultural factors, a person develops an understanding and acceptance of the specific gender roles and adapts to them.

Gender or sex role stereotypes

These are socially determined rules which contain the cultural beliefs about what the gender roles should be, as how different sexes should play separate roles according to their biological sex. Gender stereotype differs from gender role in that it tends to be the way people feel 'others' should behave.

Men are different from women. That would seem to be self-evident. They are different in aptitude skill and behavior, but then, so is every individual person. However, the stereotypes are changing as women are doing things reserved for men such as working outside home, and vice versa. Gender stereotypes, that girl are more sociable, more nurturing, and more compliant and have lower self-esteem, can no longer be sustained and over the last two decades, girls have contradicted earlier findings by showing high motivation to achieve.

It is believed that there are inherent differences between men and women on the basis of the differences in the spatial abilities between the two sexes. Men find it much easier to visualize and deal with spaces, the position of objects, relative heights and dimensions. A study also showed that in a test involving a three dimensional mechanical apparatus, only a quarter of the women could perform the task better than men. But the fact that at least some women could perform the task as well as the men is reason enough that more studies should be conducted for conclusive evidence.

In general, men are taller and heavier than women. In sports, men tend to outperform women in strength and speed. Women seem to have greater endurance. In spite of many attempts, sports have never become completely unisex. A study conducted Ward and Whipp suggested that running speeds for male and female athletes have improved steadily through the twentieth century, but women have improved much more than men. In another study by Dyer suggested the same results for athletics, swimming and cycling. Both the studies predicted that sex differences might disappear by the middle of the next century. However, Seiler and Sailer point out a temporal correlation with the use of performance enhancing drugs, and their more recent proscription. These have a greater effect on women, than they do with men, since the latter are already well supplied with testosterone.

It is also generally believed that men are more aggressive, physically and verbally, and enjoy taking risks. They play fighting games and enjoy 'dares.' More men than women are convicted for crimes, especially crimes of violence. While some psychologists call it simply a matter of biology, others suggest that it is a function of the way we organize the sex and gender roles in our society. However, the findings, in this area, have turned out to be unsatisfactory with very small differences and a large degree of overlap.

Biologically, men certainly seem to be the weaker sex because of the following reasons:

- Although one would expect there to be an equal chance of the fetus being a boy or girl, it appears that the ratio for boys is about 20 percent higher, yet only about the same number come to term. Thus there is a greater tendency for male fetuses to be aborted, miscarried or stillborn, although they have a better rate of survival after one year of birth.
- More boys are born susceptible to congenital or inherited conditions, such as hemophilia, cerebral palsy, convulsions, or heart disease.
- On average, men experience heart attacks 10 years earlier than women and symptoms also vary by sex: women experience shortness of breath, fatigue, and chest pain, while most male heart attacks come on as a sudden, striking pain in the chest.
- Moreover, in adulthood, men have greater vulnerability to virus infections and a shorter average lifespan.

In recent years, a great many biological sex differences have been found throughout the body, including the brain, both in metabolism and genetic expression. They have raised concerns about differences in the efficacy and side effects of various drugs. The current focus now is on the phenomenon of imprinting whereby a given gene from the father could silence or activate a gene from the mother, or vice versa.

However, regardless of the findings, sex differences really do exist as socially we still expect women to behave like women and men like men. Indeed the real problem is not that sex differences exist, but it is our beliefs as what sexual, or gender, behavior is appropriate or inappropriate that is a problem. These narrow and rigid concepts are often reinforced by the biological determinism argument, yet we cannot ignore the effect of learning. In fact, the belief that we are the helpless products of our heredity takes away our free will.

Differences

1. Males are more aggressive than females

Though this finding has been challenged, and the definition of aggression itself questioned, it is a fairly common feature, both of human cultures and of the more complex species that male offspring are more likely to engage in play fighting and adults more likely to fight. Many workers challenge this, while others assert that it is the primary indicator of masculinity or femininity.

2. Females have more verbal ability than males, while males have better visuo-spatial skills

The distinction seems to appear at about the age of eleven and, because of the relevance to education, it has received a great deal of attention. Although girls and boys seem to have the same ability for computational arithmetic, teenage boys also seem to do better at the more abstract math. It might seem that a childhood of social experience is the primary factor. However, the biological argument suggests that the hormonal changes of puberty activate previous dormant differences.

Thus, in nutshell it may be said that while there are inherent biological differences between men and women, but socialization and learning also play major role. As of now, the dividing line between gender roles is not so clear cut as females take on roles traditionally considered to be male territory as more and more women aspire for success at work while men help in parenting and household affairs. Interestingly, such mingling of gender roles may not remain confined to developed societies, but also in developing societies.

VIEWS AND CONSEQUENCES OF GENDER STEREOTYPING

In our society today, men and women perform distinctly different roles which are based on nothing more than their biological gender. Although these roles do not hold true for each individual, the majority of people live out their lives in accordance with these extremely pervasive roles. Sex roles can be defined as a set of behaviors and characteristics that are standard for each gender in a society. Sex role stereotypes are widely-held beliefs about those behaviors and characteristics (Singleton, 1987). The stereotypes to a large extent become the roles. Society forces people into certain roles simply by expecting that those roles are proper and enforcing them.

In general, the roles prevalent in modern Western society prescribe that men should be domineering, aggressive, and superior at the math and sciences, should become successful in their careers, and should control and suppress their feelings. Women, on the other hand, should be submissive, nurturing, gentle, better at languages and the humanities, emotional, and desirous of nothing more than a happy family and a husband to provide for her, while she remains at home and tends the house.

These sex-typed roles are perpetuated and reinforced by the mass media and society in general in many ways, some obvious and others more subtle. However, there is a previously widely-held point of view, somewhat less popular recently, that gender roles are the result of innate biological differences between the sexes; that men are naturally better-suited to hold positions of power, for example, and that women are more suited to look after the home and children.

It has been recently and frequently proposed that the currently championed roles are limiting and damaging to all involved, men and women alike, from the time that they are children. This view holds that our sex roles are solely the product of the society in which we live, and that their incompatibility with the reality of individual personalities causes pain and stress for many people, as do other forms of oppression and stereotyping. Most of the research in this area has been based on studies which point out the inequality and subordinate position of women in Western society.

The biological view of gender roles states that the differentiated gender roles which exist in our society are the products of our evolution, and are inextricably linked with abilities predominant in one gender or the other which are determined biologically. The roles prescribed for each sex are based on physical abilities and properties of that sex, such as intelligence, brain lateralization, and differing hormone levels. This view was the accepted one throughout history, and has only recently been challenged. Implicit within this view is the belief that these roles are both desirous and immutable. Ways in which this view could be substantiated include looking at gender role similarities between animals and humans, and examining sex differences among infants (Singleton, 1987). If substantial sex role similarities are found between animals and humans, or if infants are in fact found to exhibit traits traditionally assigned to their gender, this could point toward the fact that such abilities and roles are biologically determined.

The problems with this view are, first, that it assumes that current Western sex roles are the "correct" ones, second, that it denies that we can or should change our current roles, and third, that it implies that traditional sex roles are adaptive and conducive to physical and mental health. Recent evidence and research supports the conclusion that none of these three points is actually true. If current Western sex roles are in fact biologically programmed into all human beings, we would expect such roles to be universal, and this is clearly not the case. If nothing else, this view ignores, or diagnoses as pathologic, individual differences (Renzetti & Curran, 1989).

Although some consistency across many cultures is in fact found, even those who see these traits as evolutionarily based have developed other, better explanations for them than that they are biological and unalterable. One such view states that societal differences in child-rearing practices are responsible for differing abilities in each gender, but that these differences are due to the evolutionary sexual behaviour of a polygynous species (Low, 1989), as they believe humans to be (Leger, 1992).

This view, while still assuming that Western sex roles are the original ones does at least have the advantage of allowing for change over time as the species changes.

The assumption that sex roles are biological, and therefore unchanging, can be refuted by the simple observable fact that sex roles, even within our society, have changed and are in the process of changing. Women, long considered to be unable to hold positions of power or careers involving intelligence, are at last beginning to be allowed to serve in such capacities (though the struggle for complete recognition is still far from over). In addition, women are in the process of refuting the belief that they must have a family and children to be complete, when in fact many are happier without them.

As regards mental health and the adaptivity of sex roles, the simple fact is that, when a correlation between sex roles and physical and psychological health is found, it usually points to the conclusion that the female sex role in particular is highly related to lower self-esteem, higher levels of neuroticism (visible in such traits as over-sensitivity to criticism and refusal to engage in assertive behavior), and decreased ability to cope in those individuals who adhere strictly to their socially prescribed gender role. Studies have shown that women, as well as men who are considered to be highly "feminine" on the Bem Sex Role Inventory, are much more likely in situations of job stress to use avoidance coping at the expense of other, more productive, methods (Long, 1990).

In addition, women have been shown to be significantly less satisfied with their bodies, due to a gender role which states that they must be beautiful in order to attract a man, which should be of paramount importance in their lives. Even women of low body weight commonly diet, believing themselves fat. This becomes a problem when it is shown that these feelings of inadequacy about one's own body are linked to eating disorders, low self-esteem, depression, and lowered or insufficient use of contraceptives (Mintz & Betz, 1986).

Perhaps one of the reasons for the increased stress apparent among those attempting to adhere to the sex role prescribed for them (Alpert-Gillis & Connell, 1989), as well as the most compelling evidence against the biological approach, is the fact that many of the commonly-held Western gender ability stereotypes on which sex roles are based are simply inaccurate.

Researchers Maccoby and Jacklin published a review of studies of children under age two years. Children this young are frequently used because it is assumed that the societal influences on them will be negligible, and thus any differences will be due to biology (Renzetti & Curran, 1989). In their review, Maccoby and Jacklin examined three specific traits: dependency, visual-spatial ability, and activity level. If traditional Western stereotypes are in fact biological, it would be expected that the female infants would exhibit more dependency, and the male infants would exhibit greater visual-spatial ability and a higher activity level.

Instead, they found that, in regard to dependency, twelve of twenty studies showed no sex differences, and the results of the other eight were inconsistent. They also examined nine studies of newborns which showed no sex differences with regard to visual-spatial ability, and thirty-three studies of infants, in most of which visual-spatial ability was not statistically different between genders; when it was, the advantage could fall to either sex almost equally. As to activity level and aggression, no sex differences were found for infants (cited in Renzetti & Curran, 1989). Tests of spatial ability using older children show no sex differences until adolescence, when males begin to pull ahead (Renzetti & Curran, 1989).

As far as the theory that hormones or genes may control the personality traits or behaviors of each gender, so far, there has been absolutely no direct evidence that this is the case (Renzetti & Curran, 1989). Although testosterone in some animals has been shown to increase aggression, its behavior in humans is far more complicated, and shows no specific pattern of influence (Leger, 1992).

It is important to realize the inaccuracy of these gender ability stereotypes, as they often dictate such things as what careers are thought proper and what courses of action are expected for each individual. An adult who feels an affinity for something that their gender role states they should not be good at, such as a woman who wishes to take on the highly visual-spatial career of a pilot, may find herself up against both societal norms and, in the case of a military pilot, rules dictated by gender stereotypes, which prevent her from pursuing that career.

Contrary to the biological determinism perspective, the sociocultural view of gender roles states that gender roles and stereotypes develop within a culture and are then perpetuated by that culture. Individuals within the culture are expected to conform to these norms, and are socialized in manners which constantly reinforce the beliefs and behavior which are prescribed and presupposed for them.

Social learning theory applied to this area predicts that individuals need not directly be told what their expected roles are; continued exposure to others in those roles, for example in the media, is sufficient reinforcement (Long, 1990). The theory that society and not biology determines gender roles rectifies the problems inherent in the biological view. If sex roles are culturally transmitted, then Western sex roles need not, and likely are not, universal, and any roles which are adaptive within a culture are correct. This explains the findings of the Maccoby and Jacklin (1974) study (presented above) in which spatial abilities of males surpass those of females only in adolescence, by noting that during this time, socialization is more differentiated and strict, and gender socialization for women begins in earnest (Renzetti & Curran, 1989). Further supporting this view is the fact that not all societies show any sex differences on these tests. Canadian Eskimos, for example, are one such group (Renzetti & Curran, 1989).

This view is also supported by the fact that many other cultures present different gender roles from those espoused by our society. Samoans, for instance, do not differentiate social status based on gender (Mead, 1969). Native Americans, while attributing certain preferences to one sex or the other, frequently allowed for the "crossing-over" of a person to the other sex, whereby individuals of either sex (though usually males) who have the desire to do so would take on the cultural role of the other gender, marrying and laboring as their chosen gender does (Curran, 1992). These individuals (such as the *berdaches* of the plains Indians and the *nadles* of the Navajo) were generally accorded special status in certain ceremonies and social situations, but were not viewed as pathologically abnormal.

Perhaps one of the most telling studies of the phenomenon of gender roles is that done by Margaret Mead, who studied three tribes of people, all living within a twenty-mile radius of each other. One tribe, the Arapesh, socialized both their males and their females to exhibit qualities considered in our society to be "feminine": they were warm, cooperative, and nurturing, and according to their histories had always been so. The Mundugamor tribe, on the other hand, raised their children to be what we consider "masculine": competitive, aggressive, and oppositional. Once more, according to their stories, they had always been this way. The third tribe, the Tchambuli, displayed gender roles the reverse of those prevalent in our society. The women were dominant and controlling, the sexual aggressors, the principle workers, and in control. The men were emotionally dependent on the women, vain about their appearance, and reported by both themselves and the women to be irresponsible (A. S. Walters, personal communication, February 9, 1993).

Current and past changes in sex roles can moreover be explained by the fact that, since cultures change, what roles are adaptive to each culture will also change over time, and should do so. The sociocultural view, influenced to a great extent by feminism, further states that the current female sex roles in our society are psychologically damaging to women, in that they promote as desirable behaviors and beliefs which are incompatible with reality and are maladaptive to psychological adjustment. The male gender roles are also damaging to men for the same reasons.

Much of the research done on the damaging effects of gender stereotyping has focused on the way in which these stereotypes serve to further subjugate women. However, men are hurt as well. Men are told that they should never show their emotions, they are socialized to be aggressive, and they are taught to derogate anything female. This manifests itself as a high level of competitiveness, a disability to be open and vulnerable, and a lack of competence in interpersonal relationships (Kimmel, 1990).

Additionally, this directive to distrust and discount anything feminine results in men distrusting and discounting all that is feminine in themselves, denying their feminine side, and so exhibiting homophobia and misogyny, as well as a continued fear of losing their masculinity - one more stress laid onto their high-stress lifestyle, made even worse because they will never be completely free of their feminine side, and so will always feel inadequate (Starhawk, 1988).

Inherent in this denial of all things feminine is also an innate belief that masculinity and femininity are opposites. This dichotomy is detrimental to men in particular, because it teaches them that if they attempt to gain some desirable feminine qualities, they will in turn lose some of their masculinity (Starhawk, 1988), which is perhaps the ultimate terror of the gender-stereotyped male. Furthermore, both these dichotomized sex roles are damaging to society as a whole because they promote violent behavior in men, against both each other and women, discourage individuals from pursuing certain activities in which they might excel given the chance, and foster the communication gap between the sexes. For example, several studies have found that acceptance of rape myths such as "most victims are at least partly to blame," are correlated to sex role stereotyping and distrust of the opposite sex on the part of both men and women (Giacopassi & Dull, 1986).

It would seem apparent, given all the research extant, that even if it is the case that sex roles are the result of our species' evolution and the physiological predispositions of each gender, and were adaptive in the past, these roles have not altered to reflect the changing realities of our society. Therefore, any adaptive advantages they may have presented in the past are no longer present, and the roles must change in order to be adaptive for individuals today.

This prospect is certainly daunting; if gender roles are to change, then so must many other institutions of our society. The fact most if not all adults today adhere to these roles to some extent does not make this any easier. To change the outlook on women would require changes in how we perceive family relationships, how we educate our children, our criminal and civil laws, and religion, among other things (Hearn & Parkin, 1983).

Perhaps the first step to making these changes is to change the ways in which men and women are presented in the mass media. Currently, such materials as television perpetuate the traditional sex roles by presenting and emphasizing them, while discrediting those who go against the current roles by either presenting them in an unfavourable light, or, more frequently, by failing to present them at all (Durkin, 1987). The belief that all women should be young and beautiful, and that their looks should be their major concern, is perpetuated by the facts that most women on television are under thirty years old, and these women are shown constantly paying attention to their looks, and by the fact that when women do make news, such things as their marital status, height, and hair colour are frequently mentioned, even when these are irrelevant to the issue at hand (Durkin, 1987).

Such change as is necessary can be attained. In many small ways, it is already being attempted. In the DSM-III, sex bias in diagnosis of mental disorders was present in many ways. For example, in the description of histrionic behaviors, it stated that the behavior involved was "often a caricature femininity", and that histrionic individuals often act out roles such as the "princess." Also, male histrionics were thought to exhibit "homosexual arousal patterns" (DSM-III, cited in Hamilton, Rothbart, & Dawes, 1986). In the DSM-III-R, the characterization of the behaviors as feminine is no longer present, nor is the homosexual diagnosis. The battle is not completely won yet - the "princess" role reference is still present (American Psychiatric Association, 1980) - but if such changes as these can be made, a little at a time, we will be well on the way to a more enlightened view of the potential that each individual, regardless of gender, possesses.

This is an extremely important, if not necessary, realization for society to come to. Given that the current sex role stereotyping has so many negative consequences for all people in our society, and has yet to exhibit any positive effects, it stands to reason that when such an incredible force for oppression is removed from our lives, it can only benefit all involved in the long run.

Such research as has been done in the past is needed also in the future, but it must be accompanied by an active attempt to change the things that are found, rather than simply acknowledging their deleterious effects in statistical breakdowns. It must also be made clear that these roles are not universal and immutable and that there is hope for change. Only when these facts are understood can our society begin to move toward a view of gender relations that time and for the future.

2

Psychology
An Introduction

Psychology is an academic and applied discipline involving the scientific study of human mental functions and behavior. Occasionally, in addition or opposition to employing the scientific method, it also relies on symbolic interpretation and critical analysis, although these traditions have tended to be less pronounced than in other social sciences such as sociology. Psychologists study such phenomena as perception, cognition, emotion, personality, behavior and interpersonal relationships. Some, especially depth psychologists, also study the unconscious mind.

Psychological knowledge is applied to various spheres of human activity, including issues related to everyday life-such as family, education and employment-and to the treatment of mental health problems. Psychologists attempt to understand the role of mental functions in individual and social behavior, while also exploring the underlying physiological and neurological processes.

Psychology includes many sub-fields of study and applications concerned with such areas as human development, sports, health, industry, media and law. Psychology incorporates research from the natural sciences, social sciences and humanities and a person who studies or practices psychology is called a psychologist.

History

The study of psychology in philosophical context dates back to the ancient civilizations of Egypt, Greece, China and India. Psychology began adopting a more clinical and experimental approach under medieval Muslim psychologists and physicians, who built psychiatric hospitals for such purposes. In 1802, French physiologist Pierre Cabanis helped to pioneer biological psychology with his essay *Rapports du physique et du moral de l'homme* (*On the relations between the physical and moral aspects of man*). Cabanis interpreted the mind in light of his previous studies of biology, arguing that sensibility and soul are properties of the nervous system.

Though the use of psychological experimentation dates back to Alhazen's *Book of Optics* in 1021, psychology as an independent experimental field of study began in 1879, when Wilhelm Wundt founded the first laboratory dedicated exclusively to psychological research at Leipzig University in Germany, for which Wundt is known as the "father of psychology". The year 1879 is thus sometimes regarded as the "birthdate" of psychology. The American philosopher William James published his seminal book, *Principles of Psychology* in 1890, laying the foundations for many of the questions that psychologists would focus on for years to come. Other important early contributors to the field include Hermann Ebbinghaus (1850–1909), a pioneer in the experimental study of memory at the University of Berlin; and the Russian physiologist Ivan Pavlov (1849-1936) who investigated the learning process now referred to as classical conditioning.

Psychoanalysis

From the 1890s until his death in 1939, the Austrian physician Sigmund Freud developed a method of psychotherapy known as psychoanalysis. Freud's understanding of the mind was largely based on interpretive methods, introspection and clinical observations, and was focused in particular on resolving unconscious conflict, mental distress and psychopathology. Freud's theories became very well-known, largely because they tackled subjects such as sexuality, repression, and the unconscious mind as general aspects of psychological development. These were largely considered taboo subjects at the time, and Freud provided a catalyst for them to be openly discussed in polite society. While Freud is perhaps best known for his tripartite model of the mind, consisting of the id, ego, and superego, and his theories about the Oedipus complex, his most lasting legacy may be not the content of his theories but his clinical innovations, such as the method of free association and a clinical interest in dreams.

Freud had a significant influence on Swiss psychiatrist Carl Jung, whose analytical psychology became an alternative form of depth psychology. Other well-known psychoanalytic thinkers of the mid-twentieth century included Sigmund Freud's daughter, psychoanalyst Anna Freud; German-American psychologist Erik Erickson, Austrian-British psychoanalyst Melanie Klein, English psychoanalyst and physician D. W. Winnicott, German psychologist Karen Horney, German-born psychologist and philosopher Erich Fromm, and English psychiatrist John Bowlby. Contemporary psychoanalysis comprises diverse schools of thought, including ego psychology, object relations, interpersonal, Lacanian, and relational psychoanalysis. Modification of Jung's theories has led to the archetypal and process-oriented schools of psychological thought.

Austrian-British philosopher Karl Popper argued that Freud's psychoanalytic theories were presented in untestable form. Psychology departments in American universities today are scientifically oriented, and Freudian theory has been marginalized, being regarded instead as a "desiccated and dead" historical artifact, according to a recent APA study. Recently, however, South African neuroscientist Mark Solms and other researchers in the emerging field of neuropsychoanalysis have argued for Freud's theories, pointing out brain structures relating to Freudian concepts such as libido, drives, the unconscious, and repression.

Behaviorism

Behaviorism arose partly due to the popularity of laboratory-based animal experimentation and partly in reaction to Freudian psychodynamics, which was difficult to test empirically because, among other reasons, it tended to rely on case studies and clinical experience, and dealt largely with intra-psychic phenomena that were difficult to quantify or to define operationally. Moreover, in contrast with early psychologists Wilhelm Wundt and William James, who studied the mind via introspection, the behaviorists argued that the contents of the mind were not open to scientific scrutiny and that scientific psychology should only be concerned with the study of observable behavior. There was no consideration of internal representation or the mind. Founded in the early 20th century by American psychologist John B. Watson, behaviorism was embraced and extended by Americans Edward Thorndike, Clark L. Hull, Edward C. Tolman, and later B.F. Skinner.

Behaviorism differs from other perspectives in a number of ways. Behaviorists focus on behavior-environment relations and analyze overt and covert (i.e., private) behavior as a function of the organism interacting with its environment. Behaviorists do not reject the study of covert or private events (e.g., dreaming), but rather reject the proposition that an autonomous causal entity inside the organism causes overt (e.g., walking, talking) or covert (e.g., dreaming, imagining) behavior. Concepts such as "mind" or "consciousness" are not used by behaviorists because such terms do not describe actual psychological events (such as imagining) but are used as explanatory entities hidden somewhere in the organism. By contrast, behaviorism treats private events as behavior, and analyzes them in the same way as overt behavior. Behavior refers to the concrete events of the organism, overt or private.

American linguist Noam Chomsky's critique of the behaviorist model of language acquisition is regarded by many as a key turning point in the decline of behaviorism's general prominence. But Skinner's behaviorism has not died, perhaps in part because it has generated successful practical applications. The ascendancy of behaviorism as an overarching model in psychology, however, gave way to a new dominant paradigm: cognitive approaches.

Humanism and Existentialism

Humanistic psychology was developed in the 1950s in reaction to both behaviorism and psychoanalysis. By using phenomenology, intersubjectivity and first-person categories, the humanistic approach seeks to glimpse the whole person-not just the fragmented parts of the personality or cognitive functioning. Humanism focuses on uniquely human issues and fundamental issues of life, such as self-identity, death, aloneness, freedom, and meaning. There are several factors which distinguish the humanistic approach from other approaches within psychology. These include the emphasis on subjective meaning, a rejection of determinism, and a concern for positive growth rather than pathology. Some of the founding theorists behind this school of thought were American psychologists Abraham Maslow, who formulated a hierarchy of human needs, and Carl Rogers, who created and developed client-centered therapy; and German-American psychiatrist Fritz Perls, who helped create and develop Gestalt therapy. It became so influential as to be called the "third force" within psychology, along with behaviorism and psychoanalysis.

Influenced largely by the work of German philosopher Martin Heidegger and Danish philosopher Søren Kierkegaard, psychoanalytically-trained American psychologist Rollo May developed an existential breed of psychology in the 1950s and 1960s. Existential psychologists argued that people must come to terms with their mortality and that, in so doing, people will be obligated to accept that they are free-that they possess free will and are at liberty to defy expectations and conventions in order to forge their own, meaningful paths through life. May believed that an important element of the meaning-making process is the search for myths, or narrative patterns into which the individual may fit.

From the existential perspective, not only does the quest for meaning follow from an acceptance of mortality, but the attainment of meaning can overshadow the prospect of death. As Austrian existential psychiatrist and Holocaust survivor Viktor Frankl observed, "We who lived in concentration camps can remember the men who walked through the huts comforting others, giving away their last piece of bread. They may have been few in number, but they offer sufficient proof that everything can be taken from a man but one thing: the last of the human freedoms-to choose one's attitude in any given set of circumstances, to choose one's own way".

May helped to pioneer the development of existential therapy, and Frankl created a variety of it called logo therapy. In addition to May and Frankl, Swiss psychoanalyst Ludwig Binswanger and American psychologist George Kelly may be said to belong to the existential school. Both existential and humanistic psychologists argue that people should strive to reach their full potential, but only humanistic psychologists believe that this striving is innate. For existential psychologists, the striving only follows an anxiety-producing contemplation of mortality, freedom, and responsibility.

Cognitivism

Behaviorism was the dominant paradigm in American psychology throughout the first half of the 20th century. However, the modern field of psychology largely came to be dominated by cognitive psychology. Noam Chomsky's 1959 review of B. F. Skinner's *Verbal Behavior* challenged the behaviorist approaches to studies of behavior and language dominant at the time and contributed to the cognitive revolution in psychology. Chomsky was highly critical of what he considered arbitrary notions of 'stimulus', 'response' and 'reinforcement' which Skinner borrowed from animal experiments in the laboratory. Chomsky argued that Skinner's notions could only be applied to complex human behavior, such as language acquisition, in a vague and superficial manner. Chomsky emphasized that research and analysis must not ignore the contribution of the child in the acquisition of language and proposed that humans are born with a natural ability to acquire language. Work most associated with psychologist Albert Bandura, who initiated and studied social learning theory, showed that children could learn aggression from a role model through observational learning, without any change in overt behavior, and so must be accounted for by internal processes.

With the rise of computer science and artificial intelligence, analogies were drawn between information processing by humans and information processing by machines. This, combined with the assumptions that mental representations exist and that mental states and operations could be inferred through scientific experimentation in the laboratory, led to the rise of cognitivism as a popular model of the mind. Research in cognition was also backed by the aim to gain a better understanding of weapons operation since World War II.

Cognitive psychology differs from other psychological perspectives in two key ways. First, it accepts the use of the scientific method, and generally rejects introspection as a method of investigation, unlike symbol-driven approaches such as Freudian psychodynamics. Second, it explicitly acknowledges the existence of internal mental states-such as belief, desire and motivation-whereas behaviorism does not. In fact, like Freud and depth psychologists, cognitive psychologists are even interested in unconscious phenomena, including repression; but cognitive psychologists prefer to explore these phenomena in terms of operationally-defined components, such as subliminal processing and implicit memory, that are amenable to experimental investigation. Moreover, cognitive psychologists have questioned the very existence of some of these components. For example, American psychologist Elizabeth Loftus has used empirical methods to demonstrate ways in which apparent memories can be brought to light via fabrication rather than through the elimination of repression.

Preceding the cognitive revolution by several decades, Hermann Ebbinghaus had pioneered the experimental study of memory, arguing that higher mental processes are not hidden from view, but instead could be studied using experimentation. Links between psychological activity and brain and nervous system function also became understood, partly due to the experimental work of people such as English neuroscientist Charles Sherrington and Canadian psychologist Donald Hebb, and partly due to studies of people with brain injury. These mind-body links are explored at length by cognitive neurophysiologists. With the development of technologies for measuring brain function, neurophysiology and cognitive neuroscience have become increasingly active areas of contemporary psychology. Cognitive psychology has been subsumed along with other disciplines, such as philosophy of mind computer science, and neuroscience, under the umbrella discipline of cognitive science.

Schools of Thought

Various schools of thought have argued for a particular model to be used as a guiding theory by which all, or the majority, of human behavior can be explained. The popularity of these has waxed and waned over time. Some psychologists may think of themselves as adherents to a particular school of thought and reject the others, although most consider each as an approach to understanding the mind, and not necessarily as mutually exclusive theories. On the basis of Tinbergen's four questions a framework of reference of all fields of psychological research can be established (including anthropological research and humanities).

In modern times, psychology has adopted an integrated perspective towards understanding consciousness, behavior, and social interaction. This perspective is commonly referred to as the biopsychosocial approach. The basic tenet of the biopsychosocial model is that any given behavior or mental process affects and is affected by dynamically interrelated biological, psychological, and social factors.

The psychological aspect refers to the role that cognition and emotions play in any given psychological phenomenon-for example, the effect of mood or beliefs and expectations on an individual's reactions to an event. The biological aspect refers to the role of biological factors in psychological phenomena-for example, the effect of the prenatal environment on brain development and cognitive abilities, or the influence of genes on individual dispositions. The socio-cultural aspect refers to the role that social and cultural environments play in a given psychological phenomenon-for example, the role of parental or peer influence in the behaviors or characteristics of an individual.

SUBFIELDS

Psychology encompasses a vast domain, and includes many different approaches to the study of mental processes and behavior. Below are the major areas of inquiry that comprise psychology. A comprehensive list of the sub-fields and areas within psychology can be found at the list of psychology topics and list of psychology disciplines.

Abnormal Psychology

Abnormal psychology is the study of abnormal behavior in order to describe, predict, explain, and change abnormal patterns of functioning. Abnormal psychology studies the nature of psychopathology and its causes, and this knowledge is applied in clinical psychology to treat patients with psychological disorders.

It can be difficult to draw the line between normal and abnormal behaviors. In general, abnormal behaviors must be maladaptive and cause an individual significant discomfort in order to be of clinical and research interest. According to the DSM-IV-TR, behaviors may be considered abnormal if they are associated with disability, personal distress, the violation of social norms, or dysfunction. Biological psychology, Neuropsychology, Physiological psychology, and Cognitive neuroscience

MRI depicting the human brain and the arrow indicates the position of the hypothalamus. Biological psychology is the scientific study of the biological substrates of behavior and mental states. Seeing all behavior as intertwined with the nervous system, biological psychologists feel it is sensible to study how the brain functions in order to understand behavior. This is the approach taken in behavioral neuroscience, cognitive neuroscience, and neuropsychology. Neuropsychology is the branch of psychology that aims to understand how the structure and function of the brain relate to specific behavioral and psychological processes. Neuropsychology is particularly concerned with the understanding of brain injury in an attempt to work out normal psychological function. The approach of cognitive neuroscience to studying the link between brain and behavior is to use neuroimaging tools, such as to observe which areas of the brain are active during a particular task.

Cognitive psychology

Cognitive psychology studies cognition the mental processes underlying behavior. Perception, learning, problem solving, memory, attention, language and emotion are all well-researched areas. Cognitive psychology is associated with a school of thought known as cognitivism, whose adherents argue for an information processing model of mental function, informed by positivism and experimental psychology.

On a broader level, cognitive science is a conjoined enterprise of cognitive psychologists, neurobiologists' researchers in artificial intelligence, logicians, linguists, and social scientists, and places a slightly greater emphasis on computational theory and formalization. Both areas can use computational models to simulate phenomena of interest. Because mental events cannot directly be observed, computational models provide a tool for studying the functional organization of the mind. Such models give cognitive psychologists a way to study the "software" of mental processes independent of the "hardware" it runs on, be it the brain or a computer.

Comparative Psychological

Comparative psychology refers to the study of the behavior and mental life of animals other than human beings. It is related to disciplines outside of psychology that study animal behavior such as ethology. Although the field of psychology is primarily concerned with humans the behavior and mental processes of animals is also an important part of psychological research. This being either as a subject in its own right (e.g., animal cognition and ethology) or with strong emphasis about evolutionary links and somewhat more controversially as a way of gaining an insight into human psychology. This is achieved by means of comparison or via animal models of emotional and behavior systems as seen in neuroscience of psychology (e.g., affective neuroscience and social neuroscience).

Counseling

Counseling psychology seeks to facilitate personal and interpersonal functioning across the lifespan with a focus on emotional, social, vocational, educational, health-related, developmental, and organizational concerns. Counselors are primarily clinicians, using psychotherapy and other interventions in order to treat clients. Traditionally, counseling psychology has focused more on normal developmental issues and everyday stress rather than psychopathology, but this distinction has softened over time. Counseling psychologists are employed in a variety of settings, including universities, hospitals, schools, governmental organizations, businesses, private practice, and community mental health centers.

Clinical Psychology

Clinical psychology includes the study and application of psychology for the purpose of understanding, preventing, and relieving psychologically-based distress or dysfunction and to promote subjective well-being and personal development. Central to its practice are psychological assessment and psychotherapy, although clinical psychologists may also engage in research, teaching, consultation, forensic testimony, and program development and administration. Some clinical psychologists may focus on the clinical management of patients with brain injury-this area is known as clinical neuropsychology. In many countries clinical psychology is a regulated mental health profession.

The work performed by clinical psychologists tends to be done inside various therapy models, all of which involve a formal relationship between professional and client-usually an individual, couple, family, or small group-that employs a set of procedures intended to form a therapeutic alliance, explore the nature of psychological problems, and encourage new ways of thinking, feeling, or behaving. The four major perspectives are Psychodynamic, Cognitive Behavioral, Existential-Humanistic, and Systems or Family therapy. There has been a growing movement to integrate these various therapeutic approaches, especially with an increased understanding of issues regarding culture, gender, spirituality, and sexual-orientation.

With the advent of more robust research findings regarding psychotherapy, there is growing evidence that most of the major therapies are about of equal effectiveness, with the key common element being a strong therapeutic alliance. Because of this, more training programs and psychologists are now adopting an eclectic therapeutic orientation.

Critical Psychology

Critical psychology applies the methodology of critical theory to psychology. As such, it critiques not only the psychic substrates of the *status quo*, but also the elements of mainstream psychology that are, themselves, seen as contributors to oppressive ideologies. Critical psychology operates on the belief "that mainstream psychology has institutionalized a narrow view of the field's ethical mandate to promote human welfare" by promoting individual remedies to social ills, encouraging trivial and inconsequential research, and engaging in other practices that its positivistic methods fail to place under critical scrutiny.

A critical psychologist might ask whether a case of "work stress" warrants efforts to change the macro-level systems that control the work, rather than merely to treat the individual who experiences the stress-or, more accurately, who shares the stress with countless other individuals. One might also ask why "mainstream trauma efforts fail to incorporate a focus on human rights and social justice" in war-ravaged communities. In short, critical psychology seeks, where it deems appropriate, to raise psychology's level of analysis from the individual to society, and to render psychology more transformative than ameliorative. Critical psychology has been applied to a wide array of psychologies other subfields and many of its theorists are employed in mainstream psychological professions.

Developmental Psychology

Mainly focusing on the development of the human mind through the life span, developmental psychology seeks to understand how people come to perceive, understand, and act within the world and how these processes change as they age. This may focus on intellectual, cognitive, neural, social, or moral development. Researchers who study children use a number of unique research methods to make observations in natural settings or to engage them in experimental tasks. Such tasks often resemble specially designed games and activities that are both enjoyable for the child and scientifically useful, and researchers have even devised clever methods to study the mental processes of small infants. In addition to studying children, developmental psychologists also study aging and processes throughout the life span, especially at other times of rapid change (such as adolescence and old age). Developmental psychologists draw on the full range of theorists in scientific psychology to inform their research.

Educational Psychology

Educational psychology is the study of how humans learn in educational settings, the effectiveness of educational interventions, the psychology of teaching, and the social psychology of schools as organizations. The work of child psychologists such as Lev Vygotsky, Jean Piaget and Jerome Bruner has been influential in creating teaching methods and educational practices. Educational psychology is often included in teacher education programs, at least in North America, Australia, and New Zealand.

Evolutionary Psychology

Evolutionary psychology explores the genetic roots of mental and behavioral patterns, and posits that common patterns may have emerged because they were highly adaptive for humans in the environments of their evolutionary past-even if some of these patterns are maladaptive in today's environments. Fields closely related to evolutionary psychology are animal behavioral ecology, human behavioral ecology, dual inheritance theory, and sociobiology.

Memetics, founded by British evolutionary biologist Richard Dawkins, is a related but competing field that proposes that cultural evolution can occur in a Darwinian sense but independently of Mendelian mechanisms; it therefore examines the ways in which thoughts, or memes, may evolve independently of genes.

Forensic Psychology

Forensic psychology covers a broad range of practices including the clinical evaluations of defendants' reports to judges and attorneys, and courtroom testimony on given issues. Forensic psychologists are appointed by the court or hired by attorneys to conduct competency to stand trial evaluations, competency to be executed evaluations, sanity evaluations, involuntary commitment evaluations, provide sentencing recommendations, and sex offender evaluation and treatment evaluations and provide recommendations to the court through written reports and testimony. Many of the questions the court asks the forensic psychologist go ultimately to legal issues, although a psychologist cannot answer legal questions. For example, there is no definition of sanity in psychology. Rather, sanity is a legal definition that varies from place to place throughout the world. Therefore, a prime qualification of a forensic psychologist is an intimate understanding of the law, especially criminal law.

Global Psychology

Global psychology is a subfield of psychology that addresses the issues raised in the global sustainability debate. Like critical psychology, global psychology expands the objective of psychology to macro-level trends; it examines the overwhelming consequences of global warming, economic destabilization and other large-scale phenomena, while recognizing that global sustainability can best be achieved by psychologically sound individuals and cultures. Global psychologists advocate a simple and sensible, yet comprehensive, psychology, whose strength is its focus on the long-term well-being of all of humanity.

Health Psychology

Health psychology is the application of psychological theory and research to health, illness and health care. Whereas clinical psychology focuses on mental health and neurological illness, health psychology is concerned with the psychology of a much wider range of health-related behavior including healthy eating, the doctor-patient relationship, a patient's understanding of health information, and beliefs about illness. Health psychologists may be involved in public health campaigns, examining the impact of illness or health policy on quality of life and in research into the psychological impact of health and social care.

Industrial/organizational Psychology

Industrial and organizational psychology (I/O) applies psychological concepts and methods to optimize human potential in the workplace. Personnel psychology, a subfield of I/O psychology, applies the methods and principles of psychology in selecting and evaluating workers. I/O psychology's other subfield, organizational psychology, examines the effects of work environments and management styles on worker motivation, job satisfaction, and productivity.

Legal Psychology

Legal psychology is a research-oriented field populated with researchers from several different areas within psychology (although social and cognitive psychologists are typical). Legal psychologists explore such topics as jury decision-making, eyewitness memory, scientific evidence, and legal policy. The term "legal psychology" has only recently come into use, and typically refers to any non-clinical law-related research.

Occupational Health Psychology

Occupational health psychology (OHP) is a discipline that emerged out of health psychology, industrial/organizational psychology, and occupational health. OHP is concerned with identifying psychosocial characteristics of workplaces that give rise to problems in physical (e.g., cardiovascular disease) and mental health (e.g., depression). OHP has investigated such psychosocial characteristics of workplaces as workers' decision latitude and supervisors' supportiveness. OHP also concerns itself with interventions that can prevent or ameliorate work-related health problems. Such interventions have important, beneficial implications for the economic success of organizations. Other research areas of concern to OHP include workplace violence, unemployment, and workplace safety. Two exemplary OHP journals are the *Journal of Occupational Health Psychology* and *Work & Stress*. Two prominent OHP professional organizations include the European Academy of Occupational Health Psychology and the Society for Occupational Health Psychology.

Personality Psychology

Personality psychology studies enduring patterns of behavior, thought, and emotion in individuals, commonly referred to as personality. Theories of personality vary across different psychological schools and orientations. They carry different assumptions about such issues as the role of the unconscious and the importance of childhood experience. According to Freud, personality is based on the dynamic interactions of the ego, superego, and id. Trait theorists, in contrast, attempt to analyze personality in terms of a discrete number of key traits by the statistical method of factor analysis. The number of proposed traits has varied widely. An early model proposed by Hans Eysenck suggested that there are three traits that comprise human personality: extraversion-introversion, neuroticism, and psychoticism. Raymond Cattell proposed a theory of 16 personality factors. The "Big Five" or Five Factor Model, proposed by Lewis Goldberg currently has strong support among trait theorists.

Quantitative Psychology

Quantitative psychology involves the application of mathematical and statistical modeling in psychological research, and the development of statistical methods for analyzing and explaining behavioral data. The term Quantitative psychology is relatively new and little used (only recently have Ph.D. programs in quantitative psychology been formed), and it loosely covers the longer standing subfields psychometrics and mathematical psychology.

Psychometrics is the field of psychology concerned with the theory and technique of psychological measurement, which includes the measurement of knowledge, abilities, attitudes, and personality traits. Measurement of these unobservable phenomena is difficult, and much of the research and accumulated knowledge in this discipline has been developed in an attempt to properly define and quantify such phenomena. Psychometric research typically involves two major research tasks, namely: (i) the construction of instruments and procedures for measurement; and (ii) the development and refinement of theoretical approaches to measurement.

Whereas psychometrics is mainly concerned with individual differences and population structure, mathematical psychology is concerned with modeling of mental and motor processes of the average individual. Psychometrics is more associated with educational psychology, personality, and clinical psychology. Mathematical psychology is more closely related to psychonomics/experimental and cognitive, and physiological psychology and (cognitive) neuroscience.

Social Psychology

Social psychology studies the nature and causes of social behaviour. Social psychology is the study of social behavior and mental processes, with an emphasis on how humans think about each other and how they relate to each other. Social psychologists are especially interested in how people react to social situations.

They study such topics as the influence of others on an individual's behavior (e.g. conformity, persuasion), and the formation of beliefs, attitudes, and stereotypes about other people. Social cognition fuses elements of social and cognitive psychology in order to understand how people process, remember and distort social information. The study of group dynamics reveals information about the nature and potential optimization of leadership, communication, and other phenomena that emerge at least at the micro social level. In recent years, many social psychologists have become increasingly interested in implicit measures, mediational models, and the interaction of both person and social variables in accounting for behavior.

School Psychology

School psychology combines principles from educational psychology and clinical psychology to understand and treat students with learning disabilities; to foster the intellectual growth of "gifted" students; to facilitate pro social behaviors in adolescents; and otherwise to promote safe, supportive, and effective learning environments. School psychologists are trained in educational and behavioral assessment, intervention, prevention, and consultation, and many have extensive training in research. Currently, school psychology is the only field in which a professional can be called a "psychologist" without a doctoral degree, with the National Association of School Psychologists (NASP) recognizing the Specialist degree as the entry level. This is a matter of controversy as the APA does not recognize anything below a doctorate as the entry level for a psychologist. Specialist-level school psychologists, who typically receive three years of graduate training, function almost exclusively within school systems, while those at the doctoral-level are found in a number of other settings as well, including universities, hospitals, clinics, and private practice.

Research Psychology

Wilhelm Maximilian Wundt (seated) was a German psychologist, generally acknowledged as a founder of experimental psychology. Research in most areas of psychology is conducted in broad accord with the standards of the scientific method, encompassing both qualitative ethological and quantitative statistical modalities to generate and evaluate explanatory hypotheses with regard to psychological phenomena. Investigation may be pursued by experimental protocols, but alternative methods are sometimes preferred due to research ethics, the state of development in a given research domain, and other reasons.

Psychology tends to be eclectic, drawing on knowledge from other fields to help explain and understand psychological phenomena. For example, evolutionary psychologists may synthesize data from multiple subfields of psychology, biology, and anthropology. Additionally, they make extensive use of two distinctive types of reasoning. While often employing the deductive-nomological reasoning of strict positivism, they also rely on inductive reasoning to generate accounts of hunter-gatherer life that could explain the adaptive value of different thoughts and actions.

Qualitative psychological research utilizes a broad spectrum of observational methods, including action research, ethnography, exploratory statistic, structured interviews, and participant observation, to enable the gathering of rich information unattainable by classical experimentation. Research in humanistic psychology is more typically pursued via ethnographic, historical, and historiographic methods than via science. Psychodynamic research has traditionally entailed interpreting clinical case studies, and sub schools ranging from Freudian psychoanalysis to Neo-Jungian archetypal psychology have employed myth as a vehicle of interpretation. Recent developments, particularly in neuropsychoanalysis, have demanded a relatively high degree of scientific rigor.

A precursor to critical psychology, liberation psychology, cited traditional surveys in its emancipatory quests. There is debate amongst critical psychologists as to whether they should be the ones to apply the research they conduct-as to how action-oriented or awareness-oriented they should be. In general, though, their methods tend to be critical rather than positivistic, and therefore tend not only to avoid the scientific method, but also to identify the ways in which this method is improperly used and downright abused. A key concept in critical-psychological research is reflexivity, or critical self-examination that entails "a conscious exploration of how [psychologists'] own values and assumptions affect [their] theoretical and methodological goals, activities, and interpretations". Taking a reflexive approach, critical psychologists both scrutinize the current state of psychological affairs, and seek defensible positions on the "old questions-such as free will vs. determinism, nature vs. nurture, consciousness vs. unconscious forces".

The testing of different aspects of psychological function is a significant area of mainstream psychology. Psychometric and statistical methods predominate, including various well-known standardized tests as well as those created ad hoc as the situation or experiment requires.

Academic psychologists may focus purely on research and psychological theory, aiming to further psychological understanding in a particular area, while other psychologists may work in applied psychology to deploy such knowledge for immediate and practical benefit. These approaches are not mutually exclusive, and many psychologists will be involved in both researching and applying psychology at some point during their career. Many clinical psychology programs aim to develop in practicing psychologists both knowledge of and experience with research and experimental methods, which they may interpret and employ as they treat individuals with psychological issues.

When an area of interest requires specific training and specialist knowledge, especially in applied areas, psychological associations normally establish a governing body to manage training requirements. Similarly, requirements may be laid down for university degrees in psychology, so that students acquire an adequate knowledge in a number of areas. Additionally, areas of practical psychology, where psychologists offer treatment to others, may require that psychologists be licensed by government regulatory bodies as well.

Experimental Psychology

Experimental psychological research is conducted in a laboratory under controlled conditions. This method of research relies on the application of the scientific method to understand behavior. Experimenters use several types of measurements, including rate of response, reaction time, and various psychometric measurements. Experiments are designed to test specific hypotheses (deductive approach) or evaluate functional relation-ships (inductive approach).

They allow researchers to establish causal relationships between different aspects of behavior and the environment. In an experiment, one or more variables of interest are controlled by the experimenter (independent variable) and another variable is measured in response to different conditions (dependent variable). Experiments are one of the primary research methods in many areas of psychology, particularly cognitive/ psychonomics, mathematical psychology, psycho-physiology and biological psychology cognitive neuroscience.

Experiments on humans have been put under some controls, namely informed and voluntary consent. After World War II, the Nuremberg Code was established, because of Nazi abuses of experimental subjects. Later, most countries (and scientific journals) adopted the Declaration of Helsinki. In the US, the National Institute of Health established the Institutional Review Board in 1966 and in 1974 adopted the National Research Act (HR 7724). All of these measures encouraged researchers to obtain informed consent from human participants in experimental studies. A number of influential studies led to the establishment of this rule; such studies included the MIT and Fernald School radioisotope studies, the Thalidomide tragedy, the Willowbrook hepatitis study, and Stanley Milgram's studies of obedience to authority.

STATISTICAL SURVEY

Statistical surveys are used in psychology for measuring attitudes and traits, monitoring changes in mood, checking the validity of experimental manipulations, and for a wide variety of other psychological topics. Most commonly, psychologists use paper-and-pencil surveys. However, surveys are also conducted over the phone or through e-mail. Increasingly, web-based surveys are being used in research. Similar methodology is also used in applied setting, such as clinical assessment and personnel assessment.

LONGITUDINAL STUDIES

A longitudinal study is a research method which observes a particular population over time. For example, one might wish to study specific language impairment (SLI) by observing a group of individuals with the condition over a period of time. This method has the advantage of seeing how a condition can affect individuals over long time scales. However, such studies can suffer from attrition due to drop-out or death of subjects. In addition, since individual differences between members of the group are not controlled, it may be difficult to draw conclusions about the populations. Longitudinal study is a developmental research strategy that involves testing an age group repeatedly over many years. Longitudinal studies answer vital questions about how people develop. This developmental research follows people over years and the outcome has been an incredible array of findings, especially relating to psychological problems.

Naturalistic Observation

In the same way Jane Goodall studied the role of chimpanzee social and family life, psychologists conduct similar observational studies in human social, professional and family lives. Sometimes the participants are aware they are being observed and other times it is covert: the participants do not know they are being observed. Ethical guidelines need to be taken into consideration when covert observation is being carried out.

Qualitative and Descriptive Research

Research designed to answer questions about the current state of affairs such as the thoughts, feelings and behaviors of individuals is known as descriptive research. Descriptive research can be qualitative or quantitative in orientation. Qualitative research is descriptive research that is focused on observing and describing events as they occur, with the goal of capturing all of the richness of everyday behavior and with the hope of discovering and understanding phenomena that might have been missed if only more cursory examinations have been made.

Neuropsychological Methods

Neuropsychology involves the study of both healthy individuals and patients, typically who have suffered either brain injury or mental illness.

Cognitive neuropsychology and cognitive neuropsychiatry study neurological or mental impairment in an attempt to infer theories of normal mind and brain function. This typically involves looking for differences in patterns of remaining ability (known as 'functional disassociations') which can give clues as to whether abilities are comprised of smaller functions, or are controlled by a single cognitive mechanism. Artificial neural network with two layers, an interconnected group of nodes akin to the vast network of neurons in the human brain.

In addition, experimental techniques are often used to study the neuropsychology of healthy individuals. These include behavioral experiments, brain-scanning or functional neuroimaging, used to examine the activity of the brain during task performance, and techniques such as transcranial magnetic stimulation, which can safely alter the function of small brain areas to reveal their importance in mental operations.

COMPUTATIONAL MODELING

Computational modeling is a tool often used in mathematical psychology and cognitive psychology to simulate a particular behavior using a computer. This method has several advantages. Since modern computers process extremely quickly, many simulations can be run in a short time, allowing for a great deal of statistical power. Modeling also allows psychologists to visualize hypotheses about the functional organization of mental events that couldn't be directly observed in a human.

Several different types of modeling are used to study behavior. Connectionism uses neural networks to simulate the brain. Another method is symbolic modeling, which represents many different mental objects using variables and rules. Other types of modeling include dynamic systems and stochastic modeling.

Animal Studies

Animal learning experiments are important in many aspects of psychology such as investigating the biological basis of learning, memory and behavior. In the 1890s, physiologist Ivan Pavlov famously used dogs to demonstrate classical conditioning. Non-human primates, cats, dogs, rats and other rodents are often used in psychological experiments. Controlled experiments involve introducing only one variable at a time, which is why animals used for experiments are housed in laboratory settings. In contrast, human environments and genetic backgrounds vary widely, which makes it difficult to control important variables for human subjects.

CRITICISM

Criticisms of psychology often come from perceptions that it is a "fuzzy" science. Philosopher Thomas Kuhn's 1962 critique implied psychology overall was in a pre-paradigm state, lacking the agreement on overarching theory found in mature sciences such as chemistry and physics. Psychologists and philosophers have addressed the issue in various ways.

Because some areas of psychology rely on research methods such as surveys and questionnaires, critics have asserted that psychology is not scientific. Other phenomena that psychologists are interested in such as personality, thinking, and emotion cannot be directly measured and are often inferred from subjective self-reports, which may be problematic.

The validity of probability testing as a research tool has been called into question. There is concern that this statistical method may promote trivial findings as meaningful, especially when large samples are used. Some psychologists have responded with an increased use of effect size statistics, rather than sole reliance on the traditional $p<.05$ decision rule in statistical hypothesis testing.

Sometimes the debate comes from within psychology, for example between laboratory-oriented researchers and practitioners such as clinicians. In recent years, and particularly in the U.S., there has been increasing debate about the nature of therapeutic effectiveness and about the relevance of empirically examining psychotherapeutic strategies. One argument states that some therapies are based on discredited theories and are unsupported by empirical evidence. The other side points to recent research suggesting that all mainstream therapies are of about equal effectiveness, while also arguing that controlled studies often do not take into consideration real-world conditions.

FRINGE CLINICAL PRACTICES

There is also concern about a perceived gap between scientific theory and its application, in particular with the application of unproven or unsound clinical practices. Researchers such as Beyerstein (2001) say there has been a large increase in the number of mental health training programs that do not emphasize science training. According to Lilienfeld (2002) "a wide variety of invalidated and sometimes harmful psychotherapeutic methods, including facilitated communication for infantile autism... suggestive tech-niques for memory recovery (e.g., hypnotic age-regression, guided imagery, body work), energy therapies (e.g., Thought Field Therapy, Emotional Freedom Technique)... and New Age therapies of seemingly endless stripes (e.g., rebirthing, reparenting, past-life regression, Primal...therapy, neurolinguistic programming, alien abduction therapy, angel therapy) have either emerged or maintained their popularity in recent decades." Allen Neuringer made a similar point in the field of the experimental analysis of behavior in 1984.

3

Gender-Role Development

Gender-role development is one of the most important areas of human development. In fact, the sex of a newborn sets the agenda for a whole array of developmental experiences that will influence the person throughout his or her life. The often controversial study of the development of gender is a topic that is inherently interesting to parents, students, researchers, and scholars for several reasons. First and foremost, one's sex is one of the most salient characteristics that are presented to other people. Second, who one is as a male or a female becomes a significant part of one's overall identity; it is one of the first descriptors people use about them. Labeling oneself as a "boy" or "girl" can begin as early as age eighteen months. Third, gender is an important mediator of human experiences and the way in which individuals interact with each other and the physical environment. Individuals' choices of friends, toys, classes taken in middle school, and vocation all are influenced by sex. Finally, the study of sex, gender development, and sex differences becomes the focal point of an age-old controversy that has influenced the field of developmental psychology: the nature-nurture controversy. Are gender roles and sex differences biologically determined? What are the effects of society and culture on gender and sex? How do biology (nature) and environment (nurture) interact and mutually influence each other in this significant dimension of human development?

When discussing gender-role development, the definitions of the terms "sex" and "gender" need to be understood. Referring to the nature-nurture controversy, scholars have found it important to distinguish those aspects of males and females that can be attributed to biology and those that can be attributed to social influences. The term "sex" denotes the actual physical makeup of individuals that define them as male or female. Sex is determined by genetic makeup, internal reproductive organs, the organization of the brain (such as in the control of hormone production), and external genitalia. By contrast, the behavior of individuals as males or females, the types of roles they assume, and their personality characteristics, may be as much a function of social expectations and interactions as their biological makeup. For example, in American culture, females are expected to be nurturing, and males aggressive. These behaviors and characteristics are dependent upon the social context. In order to differentiate social roles and behaviors from biological features, scholars refer to these as "gender" and "gender roles." Obviously, sex and gender are intertwined. Social expectations usually are enacted once body parts reveal the biological makeup of the individual.

Both sex and gender have a developmental story to tell that begins before birth (prenatal) and continues throughout the lifespan. Important developmental changes occur from conception through the adolescence years, and there are important theoretical perspectives and research studies that have tried to shed light on these developmental accomplishments.

PRENATAL DEVELOPMENT

Gender-role development begins at conception. If the fertilized cell has an XY chromosomal pattern, the baby will become a genetic male; an XX chromosomal pattern will lead to a genetic female. There cannot be a genetic male without that Y chromosome. Sometimes there are aberrations to these patterns, which can ultimately lead to a number of syndromes such as females with only one X chromosome (Turner's syndrome) or males with two Xs and one Y (Klinefelter's syndrome). Frequently these syndromes result in some form of cognitive and physical impairment.

At around week six of gestation, the hormone testosterone will stimulate the tissues into developing into the male internal organs; otherwise, the organs will become part of the female reproductive system. Then, by around three or four months, the external genitalia are formed. It is also during early prenatal development that the brain, bathed by the male and female hormones, may differentiate into a "female" or "male" brain (for example, female brains may be more symmetrically organized), but most of this research is still inconclusive.

Prenatal sex differentiation culminates at birth. When the proclamation of "It's a boy!" or "It's a girl!" is made, the complex process of socialization begins. It is important to recognize that the path of prenatal development may take significant deviations. Aside from the chromosomal abnormalities already mentioned, there are instances during prenatal development when females are bathed by the male hormones (androgens), and situations where male genital tissues are insensitive to the differentiating function of the male hormones. Both situations can lead to a baby born with ambiguous genitalia. In such situations, parents face agonizing decisions: whether to surgically "correct" the condition and whether to raise the baby as a female or as a male.

Infancy

Overall, the sex differences between boys and girls in the first year of life are minimal. Boys may be a bit more active or fussier and girls more physically mature and less prone to physical problems, but that may be the extent of the significant differences. Yet, baby boys are bounced and roughhoused, whereas girls are talked to more. Mothers tend to ignore the emotional expressions of their infant sons, while fathers spend more time with their boys than with their girls. Even during infancy, their names, their clothing, the "sugar and spice" messages in baby congratulation cards, and their room furnishings shape girls and boys. According to Marilyn Stern and Katherine H. Karraker, adults will characterize the same baby as strong and hardy if they think it is a male, and delicate and soft if they think it is a female. In these and other ways, gender-role socialization has already begun in earnest.

Early Childhood

The years from about age two to age six are crucial years in the development of gender roles. It is during these years that children become aware of the years from about age two to age six are crucial in the development of gender roles. During these years, children become aware of their gender, where play styles and behaviors begin to crystallize around that core identity of "I am a girl" or "I am a boy." (Reflection Photolibrary/ Corbis) their gender, where play styles and behaviors begin to crystallize around that core identity of "I am a girl" or "I am a boy," and that the social context of family, school, the peer group, and the media exert potent messages in stereotyped ways. Because of the centrality of gender-role development during these years, most theories of social and personality development highlight the early childhood years. For example, in the psychoanalytic theory of Sigmund Freud, in the third stage of psychosexual development a male child encounters the Oedipal Crisis, a time when the only way in which he can cope with his desire for his mother and fear of his father is to completely identify and incorporate his father's characteristics within himself. Freud posited a similar process for girls' desires for their fathers (the Electra complex). Although many contemporary psychologists do not agree with this theory in general, Freud is credited with highlighting the development of gender and gender-role behaviors very early in childhood and their link to identification with parents.

Social learning theory, developed by Albert Bandura, emphasizes the importance of children's imitation of the behavior of others (models). The theory posits that boys learn how to behave as boys from observing and imitating masculine behaviors, especially from their fathers, and girls learn from imitating females, especially their mothers. When children imitate same-sex behaviors, they are rewarded, but imitating the other sex may carry the threat of punishment. Although the research indicates that most parents value the same behaviors for their sons and daughters, some rewards or punishments are given on the basis of gender typing, particularly during play. This is even more true for boys than for girls, with fathers being the most punitive if, for example, they observe their sons playing with Barbie dolls or sporting red fingernail polish.

Finally, cognitive developmental theory underscores the importance of understanding what it means to be a boy or girl in the development of gender roles. In 1966 Lawrence Kohlberg conceived of gender development as a three-stage process in which children first learn their identity ("I am a boy"), then gender stability ("I will always be a boy and grow up to be a man"), and finally gender constancy ("Even if I wore a dress, I would still be a boy"), all by about six years of age. A newer version of this approach, formulated by Carol Martin and Charles Halverson in 1981, emphasized the development of gender schemas- children's ideas of gender that help them categorize experiences as relevant to one sex or the other.

Regardless of which theoretical explanation of gender roles is used, the early acquisitions of such ideas and behaviors make for very stereotyped youngsters. Because young children see the world in black- and-white terms, they may go as far as to insist that only men could be physicians, even when their own pediatrician is a woman!

Middle Childhood

Whereas parents play a significant role in gender socialization when their children are very young, when most Western boys and girls enter school they separate into gender-segregated groups that seem to operate by their own set of peer-driven rules. Gender segregation is such a widespread phenomenon that boys and girls seem to work and play together only when there is a coercive adult present. During unstructured free time, the lapse into the "two cultures of childhood" (Maccoby 1998, p. 32) is quite obvious- the other sex becomes "toxic." A typical boys' group is large, competitive, hierarchical, with one or two boys at the top of the pecking order, and organized around large group outdoor activities such as sports. Rough-and-tumble play and displays of strength and toughness frequently occur. In contrast, girls' groups tend to be smaller and dependent on intense, intimate conversations where the emphasis is upon maintaining group cohesion. Girls try very hard to be "nice" to one another, even as they attempt to covertly promote their own agenda. In her 1998 book The Two Sexes, Eleanor Maccoby stated her belief that this segregation, hints of which may be seen as early as age four or five, begins when girls shy away from their exuberant, active male playmates, who do not rely as much upon language for persuasion and influence. The boys' groups ultimately evolve into a strict order that avoids anything perceived as feminine. Girls have much greater latitude in American society to cross that sacred border. Maccoby contended that these interaction styles, to some extent, continue throughout adolescence and adulthood.

Adolescence

Erik H. Erikson believed that adolescence represented a crucial turning point in the development of a sense of identity. All of the physical, social, and cognitive changes of these years lead to frequent soul-searching about "Who am I?" Such uncertainty and insecurity also can further promote conformity into one's gender role, or "gender intensification." During early adolescence, boys may emulate "macho" role models and be quite homophobic; girls may adhere to strict dress codes (e.g., that which is "in") and play down their intellectual talents and abilities. The timing of puberty may also have significant implications for adolescent gender development. Girls are more likely to encounter social difficulties when they mature early, but for boys the opposite is true.

For many adolescents, the uncertainties, conflicting demands, and withdrawal of adult and community support are predictors of significant problems. Much has been written about how difficult the adolescent years are for girls, as they are more likely than boys to experience depression, eating disorders, and low self-esteem. This may vary, however, according to the ethnicity of the girl, as African-American teenagers do not seem to express such negative views about themselves. In his 1998 book Real Boys, William Pollack emphasized the realization that gender-role socialization makes life hard for boys. Because Western culture provides boys little opportunity for self-expression and close emotional relationships, the suicide rate and rate of violence in teenage boys is far greater than for girls.

By the end of adolescence, both sexes usually become more tolerant of themselves and others in terms of their consideration of gender-related behaviors. Individuals' evolution as men and women continues throughout the lifespan, however, as each person encounters major life transitions such as marriage, parenthood, middle age, and old age. It is important to recognize that although humans emphasize the differential paths of boys and girls in the development of gender roles, the fundamental dimensions of humanity-male and female-are more similar than different.

Biological sex / Physiological sex

The biological classification of physiological bodies as male or female usually determined by external sex organs, internal sex and reproductive organs, chromosomes, hormones and secondary sexual development at puberty. Bodies with an ambiguous biological sex, i.e. with both male and female characteristics are sometimes characterised as hermaphrodite or intersex. A person's biological sex usually - but not always - corresponds with her/his gender identity.

Sexual (ity)

A category of sexual orientation, referring to person who can experience emotional, romantic, sexual and/or affectional attraction to persons of their own sex and the opposite sex. There are various forms of bisexuality, varying from being primarily attracted to persons of the same-sex and only occasionally to persons of the opposite sex (and vice versa) to being equally attracted to persons from the same sex and the opposite sex.

Feminists (Feminism)

The Safra Project defines feminists as scholars and activists challenging gender bias and/or working towards gender equality.

Gay

A term used as a synonym to homosexual. The term 'gay' is usually preferred to the term 'homosexual' when describing a person's sexual orientation, i.e. she or he "is gay". Sometimes the term gay is used to describe only male 'homosexuals'. Gay man is a man who experiences emotional, romantic, sexual and/or affectional attraction only, or primarily, to another man or other men. His sexual orientation is categorised as homosexual.

GENDER

Refers to the social and cultural codes used to distinguish between what a particular society considers 'masculine' or 'feminine' qualities, characteristics, attributes or behaviours. The definition of gender varies widely and is often the subject of exhaustive debates, although most agree that gender is largely socially and culturally determined. People are attributed a social and cultural gender that usually corresponds to their (assumed) biological sex and they are then expected to behave in accordance with gender roles as defined by their social and cultural context. Gender can be expressed in physical appearance, dress, mannerisms, speech patterns, and social behaviours and interactions. The attribution of gender to a person by others is a (unconscious) categorisation of a person as being a man or a woman. A person's own gender identity, i.e. their deeply felt (or psychological) sense of being male or female (or both, neither or other), usually - but not always - corresponds to their biological sex and to how they express their gender.

Gender Dysphoria

A medical term referring to people who have a compelling sense that their gender identity is not in conformity with the physiological or biological sex they are born with. In other words, persons who feel that they are born in the 'wrong body', i.e. biological women who feel they should have been born as men and vice versa. These persons are generally referred to as transsexual or transgender.

Gender Equality

Gender equality refers to people receiving equal opportunities to realize their full human potential according to their wish irrespective of gender. This can for example include equality in opportunities to take part in social, economic, cultural and political developments and benefiting equally from the results. It can also refer to the equality in protection of (human) rights. Gender equality does not necessarily mean 'identical conditions' or 'identical rights' as these conditions and rights may in themselves already be gendered. To achieve gender equality, recognition is needed that current social, economic, cultural, and political systems are gendered (i.e. constructed according to gender roles) which lead to women being disadvantaged in all areas of life (gender bias). This pattern is further affected by other factors of oppression and inequality such as race, ethnicity, culture, immigration status, class, age, disability, sexuality, gender identity and/or other status. Gender equality requires the empowerment of women in their particular contexts, taking their experiences and perspectives into account.

Gender Identity

Gender identity is a person's deeply felt (or psychological) sense of being male or female (or both, neither or other). A person's gender identity is the gender to which a person feels she/he belongs. This usually corresponds to a person's biological sex and to how they express their gender. However, some people have a compelling sense that their gender identity is not in conformity with their physiological or biological sex or feel and/or express a gender identity that is other than simply 'male' or 'female'. These people are usually referred to as transsexual or transgender. Gay men, lesbians and bisexual people are usually content with their gender identity and/or their biological sex, including those who are not content with their gender roles. For example those who dress or behave similar to what is socially and culturally perceived as the opposite gender, such as a woman who behaves or dresses according to what is perceived to be a 'masculine' manner.

Gender Roles

Gender affects how people perceive themselves and others and how they expect themselves and others to behave, that is, either in a 'masculine' or 'feminine' manner. These gender roles are learned and culturally and socially determined. They are also affected by factors such as education and economics. Therefore, gender roles can evolve over time. Gender roles and expectations are often identified as factors hindering gender equality. In practice gender roles usually affect women adversely in relation to many aspects of their life, such as family, socio-economic status, health, life expectation, independence, freedom and rights (gender bias).

Heterosexuality

A category of sexual orientation referring to a person who experiences emotional, romantic, sexual and/or affectional attraction only, or primarily to a person(s) of the opposite sex. This is also called 'being straight'.

Homophobia

Homophobia is an irrational fear or hatred against lesbian, gay and bisexual people and homosexuality.

Homosexuality

A category of sexual orientation referring to a person who experiences emotional, romantic, sexual and/or affectional attraction, only or primarily to a person(s) of the same sex. Persons with a homosexual orientation are also referred to as gay (both men and women) or as lesbian (women only).

Referring to a person as a 'homosexual' is usually avoided as this can be considered inappropriate or even offensive. The reasons for this are diverse and include the previous derogatory usage of the term, its medical association and the fact that 'homosexual' and 'homosexuality' in most Muslim societies and cultures refers to certain specific sexual behaviour not resembling the contemporary 'Western' understanding of homosexuality as a category of sexual orientation. Instead the Safra Project prefers using the term 'same-sex sexuality'. In addition, the term homosexual does not express the diversity of sexualities as it ignores bisexuality nor is it gender specific. When referring to persons, preference is given to the statement that someone is 'lesbian', 'gay' or 'bisexual' or that someone is 'a lesbian', 'a gay man' or 'a bisexual person' (in plural - LGB people).

Lesbian

A woman who experiences emotional, romantic, sexual and/or affectional attraction only, or primarily, to another woman or other women. Her sexual orientation is categorised as homosexual.

LGBT (Q) (people)

LGBT (Q) (people) is the acronym of lesbian, gay, bisexual and/or transgender (and/or queer) (people). It is an inclusive term for identities sometimes also associated together as 'sexual minorities'.

Polyamory

This is a new term that has emerged in the debates about non-monogamy in recent years. It literally translates into 'many loves'. Polyamory circumscribes a particular relationship philosophy that assumes that it is possible (and indeed desirable) to love many people and to maintain multiple relationships. Within polyamory there is a strong emphasis on love, intimacy, commitment and honesty.

Some have defined polyamory as 'responsible non-monogamy'. Polyfidelity is a concept closely related to polyamory. Polyfidelity is based on the understanding that the partners in a multiple relationship will be faithful towards each other, an assumption that is not clearly spelled out or implied in polyamory. While for some polyfidelity marks a very specific approach to polyamory (or even one distinct from it), others tend to equate both concepts.

Same-sex sexuality (& Same-sex relationships)

A category of sexual orientation referring to a person who experiences emotional, romantic, sexual and/or affectional attraction only or primarily, to (a) person(s) of the same sex. Same-sex relationships are emotional, romantic, sexual and/or affectional consenting relationship(s) between people of the same sex, that is, between two women or two men. Same-sex sexuality is sometimes used as a synonym for homosexuality.

However, the Safra Project normally uses the term 'same-sex sexuality' instead of, and/or distinguished from, 'homosexuality' as the notions of homosexuality that exist within most Muslim societies and cultures do not resemble the contemporary 'Western' understanding of homosexuality. In most Muslim societies and cultures, 'homosexuality' refers to certain specific sexual behaviour rather than to a category of sexual orientation that includes a wider scope of (sexual) behaviours, feelings and self-identity.

SEX

Sex refers to certain forms of sexual behaviour, i.e. 'having sex'. It can also refer to a person's biological sex, i.e. someone is from the male sex or the female sex. The term sex is sometimes confused with the term gender, just like the term biological sex is sometimes confused with gender identity. The term sex can be distinguished from sexuality and sexual orientation.

Sexual Behaviour

The factual behaviour of a person in relation to her/his sexuality, either publicly or privately, including - but not limited to – having intercourse. Sexual behaviour is different to sexual orientation as sexual behaviour refers to actions whereas sexual orientation (also) refers to feelings and to self-identity. Persons may or may not express their sexual orientation in their sexual behaviours.

Sexual Orientation

Refers to emotional, romantic, sexual and/or affectional attraction from one person to another person or persons.

Someone's sexual orientation is categorised according to the gender(s) or biological sex of the people he/she has these feelings for, that is, it describes whether a person is attracted primarily toward people of the same or the opposite sex, or to both. Sexual orientation exists along a continuum that ranges from exclusive homosexuality to exclusive heterosexuality and includes various forms of bisexuality. Sexual orientation is an important part of a person's total self-identity: how we see ourselves and how others see us. A person's experience and understanding of her/his sexual orientation can vary during their life. Sexual orientation is different from sexual behaviour because it refers to feelings and to self-identity, rather than mere actions. Persons may or may not express their sexual orientation in their behaviours. Sexual orientation can be distinguished from other aspects of sexuality such as biological sex, gender identity and gender roles.

Sexuality (Sexualities, Sexual diversity)

Sexuality refers in its broadest sense to the quality of being sexual. The term sexuality is also used in plural, i.e. 'sexualities', to reflect the diversity of sexuality, also known as 'sexual diversity'. The main aspects of sexuality are sex, biological or physiological sex, gender, gender identity, gender roles and sexual orientation.

Love is Completely Materialized

Sexuality is very important for every human being and that's why it is better to marry at young age, to develop your own feelings toward and to fulfill your duty to reproduce. Every human being needs intimate relations to feel that he/she is living and enjoying life. But men have other definitions of intimacy than women. Maybe it is better to learn more things about the female vision on intimacy and love.

Sex and pornography are typically male things in society, because men have more needs and they must prove that they are good and potent or full with brutal male energy. Maybe it is better to make it easier to find a public woman and to let young boys develop their sexual personalities, But it is even better to let young boys marry early ages and to learn how to love somebody and to take care of the weaker gender.

The biggest problems in the Netherlands: Sexual orientation. Many males have a homosexual phase in their life to experiment with the own gender. Sex is the only goal of many males here just likes a sickness that cannot be cured. Many girls do not want a friend anymore and they decide to live alone forever, without children empty lives, filled with the richness of many thousands Dutch guilders.

Having Mountains of Money yet Being Extremely Poor

Many women here told that the men are just like animals, because sex and dirty pornography (doing things) are the only goals in a intimate relation with a female. The respect and tenderness are gone in free sex societies. Only doing (mating activity) counts. The satisfaction of women is very low and relations are terminated without being started.

Why do we condemn tenderness and deep intimacy (immaterial) between a man and a woman as being old-fashioned and backward? Old traditions are based upon wisdom that is developed during many centuries. The technology of the human spirit is lost nowadays and developing it again will take many centuries. We are back at the Stone Age, where females were preys and males were the hunters.

The sex for pure physical pleasure is very destructive to the spirit of parties involved even though they may not realize it. However in a weird way this big no to premarital sex encourages the physical sex at the expense of spiritual sex. Our kids have to live two lives one to impress the Indian society (The Parents etc) that they are virtuous people. On the other hand in real life they are driven by the need and desire (after all hormones and biology is just as strong). This leads to their sexual encounter and since no one is supposed to know about it, the dilemma about different aspect of sex is never discussed with other mature people.

Studies have shown that teen pregnancy, Sexually transmitted disease, emotionally damaging sex is much more common in the group where sex is never discussed. When the parents take a strong position to denounce premarital sex no matter what it shuts of any meaningful dialogue about sex in typical *Desi* family.

Once we take the taboo away there is a room for open discussion where we might be actually help our kids engaged in destructive sex. In return, we as a society have to accept that, if two people truly love each other physical intimacy is a natural manifestation of love and how far they go is up to the two people involved. Sex is not a bad thing.

Irrespective of the social and cultural norms, pre-marital sex finds place in almost every society in some form or the other as a "natural" method of answering instinctive sexual curiosity. Ideal parenting plays an important role in explaining the phenomenon and imparting proper sexual education. Inspite of this awareness and understanding why it is still no-no to pre-marital sex? The logical reason is very obvious.

By nature sex is a pleasurable activity. Since it gives pleasure, more often this activity once committed is repeated leading to an "association" or "impressions" besides formation of habits. Those who have had pre-marital sex, unless are lucky to find a very matured and understanding partner in marriage somehow continue to suffer due to the comparison of the past with the present. The impressions and memories of the past tend to haunt thereby leading to incompatibility between the partners who in marriage have taken vows of togetherness. Obviously this is very frustrating. We see this frustration resulting in divorce or living under pressure and guilt. It is also noted that such couples are not able to understand the spiritual aspect of sex.

Pre marital Sex is Justified, Post Marital Affairs Unchecked

Leading a life of fulfillment is not merely sex but also complete understanding between the partners, strength of the marriage institution and also raising of a loving and harmonious family. By all logical reasoning, it is not possible to achieve this sense of fulfillment if the mind is grossly engaged in either pre-marital or post-marital or extra-marital affairs for it means dissipitation of energies and leading an unfocussed life.

Thus no-no to pre-marital sex is not a taboo but an ideal step towards leading and living life with a sense of purpose and fulfillment.

TRANSGENDER (TRANSGENDER PEOPLE)

Some use transgender/transgender people as a synonym for transsexual(s) or to refer to persons medically diagnosed with gender dysphoria. Others use the term 'transgender' more widely to refer to all expressions of gender identity other than simply 'male' or 'female'. Transgender can vary from a surgical change to someone's biological sex (transsexual (ity)) to occasionally dressing in the clothing of the 'opposite' gender (transvestite). A transgender person can also transition - sometimes with the help of hormone therapy and/or cosmetic surgery - to live in the gender role of choice, without undergoing or wishing to undergo (complete) surgery. The term transgender can also include those who identify and/or express themselves as neither male nor female and/or those born with an ambiguous biological sex.

The International Foundation for Gender Education defines a transgender person as "someone whose gender display at least sometimes runs contrary to what other people from the same culture would normally expect." Female to male (FTM) transgender people are born with female bodies but have a predominantly male gender identity, male to female (MTF) transgender people were born with male bodies but have a predominantly female gender identity. The Safra Project uses the widest possible definition of transgender and includes those who feel not able to express or display the gender identity of their choice, for example for fear of negative reactions.

Transphobia

An irrational fear or hatred against transgender people.

Transsexuality

Refers to people who have a compelling sense that their gender identity is not in conformity with the physiological or biological sex they are born with. This may lead some to seek gender (or sex) reassignment surgery to make her/his biological or physiological sex correspond to her/his gender identity. Some people include under the term transsexual only persons who have already undergone (complete) gender (or sex) reassignment surgery, others include those who wish to undergo gender (or sex) reassignment surgery. Some also define the term more widely and use it partially or completely as a synonym for transgender. As the Safra Project uses the term transgender in the widest possible sense, it usually refers only to those having undergone gender (or sex) reassignment surgery when using the term transsexual but respects the rights of others to self-identify as transsexual.

Transvestite (Cross dresser)

A person who dresses in the clothing of the 'opposite' gender, either occasionally or always. Generally, these persons do not wish to alter the biological (or psysiological) sex of their body.

Queer

A term often used as a slur in English to refer to lesbian, gay, bisexual, transgender and queer (LGBTQ) persons. The term has been reclaimed by many LGBTQ people, often in a activist or political context, as an expression of pride in sexual diversity and variations of gender identity. The word queer has now become a commonly used term in social science studies and many universities offer for example courses in Queer Studies.

GENDER ROLES

Gender roles can be defined as the behaviors and attitudes expected of male and female members of a society by that society. Gender roles vary. Different cultures impose different expectations upon the men and women who live in that culture. The United States has experienced tremendous upheaval and revising of its traditional gender roles in the last generation. These changes in gender roles affect the home, the workplace, and the school, and they affect all Americans to some degree.

Gender Roles in the Workplace

Over the past few decades, Americans have made great strides in accepting and adjusting to new definitions of gender roles. Part of the cause is the increased number of women in the workplace. According to the Bureau of Labor Statistics, white men (who once dominated the workplace) now account for about 45percent of all workers. White women and women of color make up 47 percent of the workplace. In 1995, 76 percent of women between the ages of 25 and 54 worked outside the home, up from 50 percent in 1970.

With the increased presence of women in the workplace, old attitudes and behaviors have had to change. Men and women are more aware of sexual harassment than previously; whereas 20 years ago a woman who refused to have an affair with her boss may have had to quit, she now has other options. Companies are now experimenting with policies that are family-friendly, such as flex time, job sharing, and on-site child care-policies that benefit both men and women.

In the nascently and experimentally egalitarian workplace, some men are concerned about being accused of sexual harassment, and they feel they must be extremely cautious in their everyday dealings with the women they work with; this caution may stifle creativity, some experts say. In addition, women still earn far less money than men do for the same work, even though their salaries are vital to maintaining their families' economic health.

Differences between the Sexes

Physical differences do exist between males and females. Studies of the brain have revealed that female brains are stronger in the left hemisphere, which rules language. As a result, they do better when tested for language ability and speech articulation, for example. In males, the right hemisphere, which governs spatial perception, is stronger, giving them an advantage in tasks that require moving objects or aiming.

Tasks or tests that do not take into account the differences between males and females tend to penalize one gender or the other. For example, boys tend to score better on standardized achievement tests, but girls do better on tests that require writing. Math, science, and geography are subjects that males tend to do better in than females, but females may have the advantage in meeting the social expectations of school, such as behaving in class and producing neat work. Even if a boy and a girl were raised identically, without gender expectations, they would not turn out the same, researchers say.

Boys and Gender Roles

Research into the differences between girls and boys is relatively new and is politically charged. Some researchers fear being labeled "anti-female" by delving into the study of boys and gender roles. But over the last three decades, gender roles have changed dramatically, and the impact on boys needs to be examined.

Some researchers maintain that boys may not develop their full capacity for emotional depth because of a combination of factors, including parenting, education, biological and genetic factors, and the messages they receive from popular culture. As a result, some boys are less able than girls to deal with the emotional upheavals that accompany adolescence; recent statistics show that teen age boys commit suicide at five times the rate teenage girls do. Ultimately, a lack of emotional development as a boy makes it difficult for the adult man to develop healthy relationships.

As gender roles have changed, they have opened greater opportunities for females (which will be discussed more fully below). But men face a dilemma. The old model of the "macho man" is less acceptable in today's world than it was even three decades ago, and men are struggling to reinvent themselves. Some men are so dependent on the old roles for their identity that they find themselves at a loss when confronted with new expectations. For example, some men cannot adjust when they discover that their wives or girlfriends earn more money than they do, and end the relationship. To some, perhaps. But plainly, for such men the new options they have regarding gender roles are limited and limiting.

What does it mean to be a man? That's a question many of today's men are wrestling with. In his book Reaching up for Manhood, author Geoffrey Canada wrote, "The image of male as strong is mixed with the image of male as violent. Male as virile gets mixed with male as promiscuous. Males as intelligent often get mixed with male as arrogant, racist, and sexist." Small wonder that so many men in western society are flailing about for a new definition. However, today's parents have the opportunity to show their sons that they don't have to be violent to be strong. Rather than taking the attitude that "boys will be boys" if their son gets into a fight, parents can take the chance to teach their child new ways to solve conflicts-without using fists.

Women and Gender Roles

Just as men's gender roles have changed, women's gender roles have changed in the last few years, opening new opportunities. However, opportunities have their price, and some things are slower to change than others.

Women can no longer be discriminated against in the workplace. If a woman is qualified for a job, she is by law able to have it. However, few women hold top positions at large companies. A 1995 survey found that among *Fortune* 500 companies, only 90 had women as their chief executive officers. About 65 percent of Americans believe that women are discriminated against in getting such well-paying positions-a phenomenon called the "glass ceiling," in which a woman rises only so far in management and no further.

However, women are looking more and more at the tradeoffs involved. Even though they may be able to get ahead in the workplace, things at home remain remarkably the same as they did in their parents' generation. Due to gender roles, women-even if they work full-time outside the home-are still perceived as having the primary responsibility for taking care of home and family. Generally, if a child is sick and both parents work, it is the mother who leaves the office, picks the child up, and stays home until the child is well enough to return to school. Researchers have also found that the woman is still the primary doer of housework (although today's men tend to do more housework than their fathers did). Working mothers do 20 hours of housework each week, compared to working fathers, who do 10. The tasks considered "male," such as yard work and car maintenance, were sporadic in nature and involved an aspect of leisure. The tasks considered "female" were generally repetitive, and had to be done daily-researchers called these tasks "unrelenting, repetitive, and routine." In addition, women are still responsible for most of the food shopping, child care, laundry, cleaning, cooking, and even for how the house looks.

Despite great changes in the workplace, life at home is still much the same as it was in past generations. Women do most of the work. Men earn most of the money. And this is not sitting well with women: researchers report that 38 percent have a problem with how much their husbands do. It's a dilemma for women to feel they have the right to choose their own career paths, an opportunity few of their mothers and grandmothers had, and then realize that they are still ruled by many of the same old gender role expectations.

The More Things Change

The shifting of gender roles in the past 30 years has been huge. It has happened so quickly that men and women are still trying to sort out what the new roles and rules mean to them. Although women are no longer expected to be the keepers of the house, in reality, they are in most families. Although men are generally open to the successes enjoyed by the women they share their lives with, some still find it hard to celebrate a woman's triumphs because they feel it diminishes their own.

However, rather than blaming each other for the situation, men and women are increasingly willing to work together to learn about their new roles. Successful marriage partners learn to negotiate and share tasks. Managers take employees aside and tell them when comments are inappropriate. It will take time to sort out all the implications of the changing gender roles of Americans, but new expectations should result in better workplaces, better relationships, better schools, and better lives.

GENDER ROLES AND ISLAM

Islam can mean many different things to different people. Sometimes people talk about 'Islam' when referring to the culture or traditions from a particular country or from a specific group of people. Sometimes people use the word 'Islam' to refer to the practice of religious rituals and/or to spirituality. People also use the term 'Islam' to talk about a political viewpoint and sometimes they are referring to what is known as 'Islamic law' or shari'ah. This body of rules, norms and laws is itself made up of several schools of thought and differing individual opinions of Muslim scholars. The Safra Project uses the term 'Muslim laws' to refer to both shari'ah as well as to modern state laws said to be based on it.

In Muslim societies women and men are expected to behave in line with social, cultural or religious codes (gender roles). These are created to distinguish between what is considered to be 'masculine' or 'feminine' (gender). These gender roles are learned within a particular social and cultural context and are affected by education and economics. In practice gender roles often affect women adversely impeding their self determination in areas like their socio-economic status, status within the family, health, life expectation, independence, freedom and rights (gender bias).

Although the Qur'an views women and men to be equal in human dignity, this spiritual or ethical equality has not been reflected in most Muslim laws. For example, women do not have equal rights to make independent decisions about choice of (marriage) partner, getting a divorce and custody of their children. Reformists and feminists have challenged women's lack of rights and lack of control over their own lives in Muslim laws through the various techniques as in the section 'Framework for Progressive Islam'.

Central to this challenge has been the reinterpretation of Qur'anic verses which seemingly privilege men over women and reinforce gender roles. Qur'an verse 4.34 refers to men as 'guardians' (qawamun) (over women), has been used to justify gender roles and male privilege over women. (For details of this and other verses see Wadud, Hassan & Mernissi). Reformist and feminist scholars have argued that the concept of guardianship has formed the basis of particular 'gendered' roles in Muslim societies. Women are often expected to be obedient wives and mothers staying within the family environment and men are expected to be protectors and caretakers of the family (Hassan, Wadud, Barlas).

These scholars have explored how verse 4.34 has been interpreted and used to limit women's autonomy, freedom of movement and access to economic opportunities and independence. They believe that the concept of 'guardianship' only meant to make sure that a woman who is bearing and nurturing children, is provided for (by her husband) whilst undertaking this task. Feminist scholars believe that this economic safeguard has been extended through the concept of guardianship to create a rigid division of gender roles and social control of men over women (Hassan, Wadud, Yamani). This extension of male 'guardianship' over women has become embodied in Muslim laws and is embedded in Muslim societies.

One of the key reasons suggested for justifying male guardianship over women within the family and in society at large, is the idea that female sexuality needs to be controlled. The concept of guardianship, rigid gender roles and male control over women's sexuality are also tools to impose and enforce heterosexuality.

Shari'ah

A body of rules, norms and laws according to which Muslims (are supposed to) live their lives. These rules, norms and laws are found in, and derived from two main sources:

the Qur'an and practices and sayings of the Prophet Muhammad (hadith). The Qur'an and hadith are known as the primary sources of Muslim law Shar'iah was formulated between the eighth and the fourteenth century AD. In the eight and ninth century AD several Muslim schools of thought (madahibs) emerged in different geographical locations, deriving legal and religious rules from the Qur'an and hadith. These schools also formulated (legal) opinions known as jurisprudence (fiqh), including the assessment of the reliability of hadith. For situations that were not explicitly mentioned in the Qur'an or hadith, the schools used certain methods of legal reasoning to formulate rules. These methods include the formulation of a consensus opinion (ijma), drawing an analogy (qiyas) from an existing rule or law or making a decision on the basis of (principles of) social justice (maslaha).

Differing opinions between the schools as well as between individual scholars, combined with influences of local customs and regional differences, caused variations in the rules, norms and laws within and between the schools. This body of rules, opinions and laws, is referred to as 'classical Muslim law' or shari'ah. Four Sunni schools (Hanafi, Hanbali, Maliki and Shafi'i) and one Shi'a (Jafari) school remain dominant today. The process of understanding and formulating legal and religious opinions from the Qur'an and hadith is known as interpretation (ijtihad).It is believed by traditional Muslim scholars that somewhere between the tenth and the fourteenth century 'the gate of ijtihad' was closed, preventing new interpretations of the Qur'an being recognised as shari'ah. Ijtihad was then replaced by the doctrine of taqlid or imitation of the rules, norms and laws that already existed. The idea of taqlid has made it difficult for reformists to challenge the shari'ah and formulate new interpretations of the Qur'an.

Many countries with significant Muslim populations have developed so-called 'shari'ah laws' as state law, particularly in the area of 'personal status' or family laws. These are either based on, or are variations of, the jurisprudence of one or more of the five schools of thought. The Safra Project uses the term

'Muslim laws' to refer to both shari'ah as well as current state laws that are said to be based on shari'ah.

SEX ROLES

In our society today, men and women perform distinctly different roles which are based on nothing more than their biological gender. Although these roles do not hold true for each individual, the majority of people live out their lives in accordance with these extremely pervasive roles. Sex roles can be defined as a set of behaviors and characteristics that are standard for each gender in a society. Sex role stereotypes are widely-held beliefs about those behaviors and characteristics (Singleton, 1987). The stereotypes to a large extent become the roles. Society forces people into certain roles simply by expecting that those roles are proper and enforcing them.

In general, the roles prevalent in modern Western society prescribe that men should be domineering, aggressive, and superior at the maths and sciences, should become successful in their careers, and should control and suppress their feelings. Women, on the other hand, should be submissive, nurturing, gentle, better at languages and the humanities, emotional, and desirous of nothing more than a happy family and a husband to provide for her, while she remains at home and tends the house.

These sex-typed roles are perpetuated and reinforced by the mass media and society in general in many ways, some obvious and others more subtle. However, there is a previously widely-held point of view, somewhat less popular recently, that gender roles are the result of innate biological differences between the sexes; that men are naturally better-suited to hold positions of power, for example, and that women are more suited to look after the home and children.

It has been recently and frequently proposed that the currently championed roles are limiting and damaging to all involved, men and women alike, from the time that they are children. This view holds that our sex roles are solely the product of the society in which we live, and that their incompatibility with the reality of individual personalities causes pain and stress for many people, as do other forms of oppression and stereotyping. Most of the research in this area has been based on studies which point out the inequality and subordinate position of women in Western society.

The biological view of gender roles states that the differentiated gender roles which exist in our society are the products of our evolution, and are inextricably linked with abilities predominant in one gender or the other which are determined biologically. The roles prescribed for each sex are based on physical abilities and properties of that sex, such as intelligence, brain lateralization, and differing hormone levels. This view was the accepted one throughout history, and has only recently been challenged. Implicit within this view is the belief that these roles are both desirous and immutable. Ways in which this view could be substantiated include looking at gender role similarities between animals and humans, and examining sex differences among infants (Singleton, 1987). If substantial sex role similarities are found between animals and humans, or if infants are in fact found to exhibit traits traditionally assigned to their gender, this could point toward the fact that such abilities and roles are biologically determined.

The problems with this view are, first, that it assumes that current Western sex roles are the "correct" ones, second, that it denies that we can or should change our current roles, and third, that it implies that traditional sex roles are adaptive and conducive to physical and mental health. Recent evidence and research supports the conclusion that none of these three points is actually true. If current Western sex roles are in fact biologically programmed into all human beings, we would expect such roles to be universal, and this is clearly not the case. If nothing else, this view ignores, or diagnoses as pathologic, individual differences (Renzetti & Curran, 1989).

Although some consistency across many cultures is in fact found, even those who see these traits as evolutionarily based have developed other, better explanations for them than that they are biological and unalterable.

The assumption that sex roles are biological, and therefore unchanging, can be refuted by the simple observable fact that sex roles, even within our society, have changed and are in the process of changing. Women, long considered to be unable to hold positions of power or careers involving intelligence, are at last beginning to be allowed to serve in such capacities (though the struggle for complete recognition is still far from over). In addition, women are in the process of refuting the belief that they must have a family and children to be complete, when in fact many are happier without them.

As regards mental health and the adaptivity of sex roles, the simple fact is that, when a correlation between sex roles and physical and psychological health is found, it usually points to the conclusion that the female sex role in particular is highly related to lower self-esteem, higher levels of neuroticism (visible in such traits as over-sensitivity to criticism and refusal to engage in assertive behavior), and decreased ability to cope in those individuals who adhere strictly to their socially prescribed gender role. Studies have shown that women, as well as men who are considered to be highly "feminine" on the Bem Sex Role Inventory, are much more likely in situations of job stress to use avoidance coping at the expense of other, more productive, methods (Long, 1990).

In addition, women have been shown to be significantly less satisfied with their bodies, due to a gender role which states that they must be beautiful in order to attract a man, which should be of paramount importance in their lives. Even women of low body weight commonly diet, believing themselves fat. This becomes a problem when it is shown that these feelings of inadequacy about one's own body are linked to eating disorders, low self-esteem, depression, and lowered or insufficient use of contraceptives (Mintz & Betz, 1986).

Perhaps one of the reasons for the increased stress apparent among those attempting to adhere to the sex role prescribed for them (Alpert-Gillis & Connell, 1989), as well as the most compelling evidence against the biological approach, is the fact that many of the commonly-held Western gender ability stereotypes on which sex roles are based are simply inaccurate.

Researchers Maccoby and Jacklin published a review of studies of children under age two years. Children this young are frequently used because it is assumed that the societal influences on them will be negligible, and thus any differences will be due to biology (Renzetti & Curran, 1989). In their review, Maccoby and Jacklin examined three specific traits: dependency, visual-spatial ability, and activity level. If traditional Western stereotypes are in fact biological, it would be expected that the female infants would exhibit more dependency, and the male infants would exhibit greater visual-spatial ability and a higher activity level.

Instead, they found that, in regard to dependency, twelve of twenty studies showed no sex differences, and the results of the other eight were inconsistent. They also examined nine studies of newborns which showed no sex differences with regard to visual-spatial ability, and thirty-three studies of infants, in most of which visual-spatial ability was not statistically different between genders; when it was, the advantage could fall to either sex almost equally. As to activity level and aggression, no sex differences were found for infants (cited in Renzetti & Curran, 1989). Tests of spatial ability using older children show no sex differences until adolescence, when males begin to pull ahead (Renzetti & Curran, 1989).

As far as the theory that hormones or genes may control the personality traits or behaviors of each gender, so far, there has been absolutely no direct evidence that this is the case (Renzetti & Curran, 1989). Although testosterone in some animals has been shown to increase aggression, its behavior in humans is far more complicated, and shows no specific pattern of influence (Leger, 1992).

It is important to realize the inaccuracy of these gender ability stereotypes, as they often dictate such things as what careers are thought proper and what courses of action are expected for each individual. An adult who feels an affinity for something that their gender role states they should not be good at, such as a woman who wishes to take on the highly visual-spatial career of a pilot, may find herself up against both societal norms and, in the case of a military pilot, rules dictated by gender stereotypes, which prevent her from pursuing that career.

Contrary to the biological determinism perspective, the socio-cultural view of gender roles states that gender roles and stereotypes develop within a culture and are then perpetuated by that culture. Individuals within the culture are expected to conform to these norms, and are socialized in manners which constantly reinforce the beliefs and behavior which are prescribed and presupposed for them.

Social learning theory applied to this area predicts that individuals need not directly be told what their expected roles are; continued exposure to others in those roles, for example in the media, is sufficient reinforcement (Long, 1990). The theory that society and not biology determines gender roles rectifies the problems inherent in the biological view. If sex roles are culturally transmitted, then Western sex roles need not, and likely are not, universal, and any roles which are adaptive within a culture are correct. This explains the findings of the Maccoby and Jacklin (1974) study (presented above) in which spatial abilities of males surpass those of females only in adolescence, by noting that during this time, socialization is more differentiated and strict, and gender socialization for women begins in earnest (Renzetti & Curran, 1989). Further supporting this view is the fact that not all societies show any sex differences on these tests. Canadian Eskimos, for example, are one such group (Renzetti & Curran, 1989).

This view is also supported by the fact that many other cultures present different gender roles from those espoused by our society. Samoans, for instance, do not differentiate social status based on gender (Mead, 1969). Native Americans, while attributing certain preferences to one sex or the other, frequently allowed for the "crossing-over" of a person to the other sex, whereby individuals of either sex (though usually males) who have the desire to do so would take on the cultural role of the other gender, marrying and laboring as their chosen gender does (Curran, 1992). These individuals (such as the *berdaches* of the plains Indians and the *nadles* of the Navajo) were generally accorded special status in certain ceremonies and social situations, but were not viewed as pathologically abnormal.

Perhaps one of the most telling studies of the phenomenon of gender roles is that done by Margaret Mead, who studied three tribes of people, all living within a twenty-mile radius of each other. One tribe, the Arapesh, socialized both their males and their females to exhibit qualities considered in our society to be "feminine": they were warm, cooperative, and nurturing, and according to their histories had always been so. The Mundugamor tribe, on the other hand, raised their children to be what we consider "masculine": competitive, aggressive, and oppositional. Once more, according to their stories, they had always been this way. The third tribe, the Tchambuli, displayed gender roles the reverse of those prevalent in our society. The women were dominant and controlling, the sexual aggressors, the principle workers, and in control. The men were emotionally dependent on the women, vain about their appearance, and reported by both themselves and the women to be irresponsible (A. S. Walters, personal communication, February 9, 1993).

Current and past changes in sex roles can moreover be explained by the fact that, since cultures change, what roles are adaptive to each culture will also change over time, and should do so. The sociocultural view, influenced to a great extent by feminism, further states that the current female sex roles in our society are psychologically damaging to women, in that they promote as desirable behaviors and beliefs which are incompatible with reality and are maladaptive to psychological adjustment. The male gender roles are also damaging to men for the same reasons.

Much of the research done on the damaging effects of gender stereotyping has focused on the way in which these stereotypes serve to further subjugate women. However, men are hurt as well. Men are told that they should never show their emotions, they are socialized to be aggressive, and they are taught to derogate anything female. This manifests itself as a high level of competitiveness, a disability to be open and vulnerable, and a lack of competence in interpersonal relationships (Kimmel, 1990).

Additionally, this directive to distrust and discount anything feminine results in men distrusting and discounting all that is feminine in themselves, denying their feminine side, and so exhibiting homophobia and misogyny, as well as a continued fear of losing their masculinity - one more stress laid onto their high-stress lifestyle, made even worse because they will never be completely free of their feminine side, and so will always feel inadequate (Starhawk, 1988).

Inherent in this denial of all things feminine is also an innate belief that masculinity and femininity are opposites. This dichotomy is detrimental to men in particular, because it teaches them that if they attempt to gain some desirable feminine qualities, they will in turn lose some of their masculinity (Starhawk, 1988), which is perhaps the ultimate terror of the gender-stereotyped male. Furthermore, both these dichotomized sex roles are damaging to society as a whole because they promote violent behavior in men, against both each other and women, discourage individuals from pursuing certain activities in which they might excel given the chance, and foster the communication gap between the sexes. For example, several studies have found that acceptance of rape myths such as "most victims are at least partly to blame," are correlated to sex role stereotyping and distrust of the opposite sex on the part of both men and women (Giacopassi & Dull, 1986).

It would seem apparent, given all the research extant, that even if it is the case that sex roles are the result of our species' evolution and the physiological predispositions of each gender, and were adaptive in the past, these roles have not altered to reflect the changing realities of our society. Therefore, any adaptive advantages they may have presented in the past are no longer present, and the roles must change in order to be adaptive for individuals today.

This prospect is certainly daunting; if gender roles are to change, then so must many other institutions of our society.

The fact that most if not all adults today adhere to these roles to some extent does not make this any easier. To change the outlook on women would require changes in how we perceive family relationships, how we educate our children, our criminal and civil laws, and religion, among other things (Hearn & Parkin, 1983).

Perhaps the first step to making these changes is to change the ways in which men and women are presented in the mass media. Currently, such materials as television perpetuate the traditional sex roles by presenting and emphasizing them, while discrediting those who go against the current roles by either presenting them in an unfavourable light, or, more frequently, by failing to present them at all (Durkin, 1987). The belief that all women should be young and beautiful, and that their looks should be their major concern, is perpetuated by the facts that most women on television are under thirty years old, and these women are shown constantly paying attention to their looks, and by the fact that when women do make news, such things as their marital status, height, and hair colour are frequently mentioned, even when these are irrelevant to the issue at hand (Durkin, 1987).

Such change as is necessary can be attained. In many small ways, it is already being attempted. In the DSM-III, sex bias in diagnosis of mental disorders was present in many ways. For example, in the description of histrionic behaviors, it stated that the behavior involved was "often a caricature femininity", and that histrionic individuals often act out roles such as the "princess." Also, male histrionics were thought to exhibit "homosexual arousal patterns" (DSM-III, cited in Hamilton, Rothbart, & Dawes, 1986). In the DSM-III-R, the characterization of the behaviors as feminine is no longer present, nor is the homosexual diagnosis. The battle is not completely won yet - the "princess" role reference is still present (American Psychiatric Association, 1980) - but if such changes as these can be made, a little at a time, we will be well on the way to a more enlightened view of the potential that each individual, regardless of gender, possesses.

This is an extremely important, if not necessary, realization for society to come to. Given that the current sex role stereotyping has so many negative consequences for all people in our society, and has yet to exhibit any positive effects, it stands to reason that when such an incredible force for oppression is removed from our lives, it can only benefit all involved in the long run. Such research as has been done in the past is needed also in the future, but it must be accompanied by an active attempt to change the things that are found, rather than simply acknowledging their deleterious effects in statistical breakdowns. It must also be made clear that these roles are not universal and immutable and that there is hope for change. Only when these facts are understood can our society begin to move toward a view of gender relations that is adaptive for our time and for the future.

PRE-MARITAL SEX AND INDIAN CULTURE

Pre-marital sex is a big no-no in our çulture for various reasons. While prohibiting sex until one is married has its own merits, one must also consider the fact that in recent times the age in which the men and women are tying the nuptial knots has been slowly on the rise. Early 30s is a common age until which men and women remain un-married due to prevalent socio-economic situation. In light of this, if Indian values have undergone any changes or do Indians still consider pre-marital sex a sin. Please share your thoughts and views.

Some Responses

Pre-marital sex has many forms...from reading novels, stories, watching films (including porno) and also expressing through poems, writings and expressions through sculpture and paintings etc. Even jokes and humor can be counted in this category where dual meaning dialogs seem to fulfill the insatiable desire. True social norms prohibit pre-marital sex and for valid reasons, yet the baser instinct always prevails and people are able to find outlets in various forms. Some may even find refuge in singing devotionals in temples (don't they get an opportunity to ogle the opposite sex!). Thus Taboo or no taboo.... the fact remains that each one has to find his/her own way to meet the urges arising what you address as pre-marital sex.

Sex is a very strong biological need. However sex after marriage has a purpose of not only evolution but also a medium of expressing love between couples. If sex becomes a mere need of the body then it can not be enjoyed in true sense and becomes monotonous. This is the reason why many marriages are failing now a day. Because the concept of sex is not very clear and a lot of importance is paid to this subject in negative sense. It is therefore duty of the parents to talk freely on the subject with their kids and enlighten them. The concepts of pre-marital sex have a direct bearing on our total concepts of sex itself. It is important how one feels about sex personally, it is like when you are hungry, you eat garbage or you wait for the right food. Or in other words reading a good book.

Not only is the act of sex, but also discussions and conversations about sex, specifically premarital sex is taboo in our society. As difficult as, it is a parent's responsibility to inform their children about sexual matters, including diseases, birth control, responsibility and the limits that they see necessary. After those frank and open discussions between parents and children, it then becomes the child's responsibility to act wisely in a situation. As for an age, I personally think that sexual activity should be curtailed until the child becomes an adult at 18. At this time they need to be held responsible for their own actions.

Different opinions of Various People

1. Reason

Good for the following reasons:

1. Helps one evaluate one's own sexuality.
2. Helps one understand the opposite sex at a deeper level.

Bad for the following reasons:

1. Emotional account dissipates (ref: well known "village theory")
2. When misunderstood (highly likely), one tend to resort to junk food as someone already mentioned in this thread.
3. If you don't behave as Romans do in Rome, you can not exist peacefully (sex is not the only thing one needs in life).

4

Gender Stress

Causes of female and male depression and anxiety might be quite different from each other. It is not known for sure if stress affects men and women differently. Generally, as the two genders often operate in different social contexts, both tend to develop different emotional dispositions and personality traits. Accordingly, their responses and coping mechanisms to stress situations vary.

Depression is one of the most common New Age maladies. All of us, at one point or the other has been down in the dumps. And it's not surprising to know that, Man has been plagued by depression since time immemorial. Various remedies have also existed for curing the blues. So maybe it might be a good idea to go back and look at some of these remedies.

INTRODUCTION

The word 'stress' is defined by the Oxford Dictionary as "a state of affair involving demand on physical or mental energy". A condition or circumstance (not always adverse), which can disturb the normal physiological and psychological functioning of an individual. In medical parlance 'stress' is defined as a perturbation of the body's homeostasis. This demand on mind-body occurs when it tries to cope with incessant changes in life. A 'stress' condition seems 'relative' in nature. Extreme stress conditions, psychologists say, are detrimental to human health but in moderation stress is normal and, in many cases, proves useful. Stress, nonetheless, is synonymous with negative conditions. Today, with the rapid diversification of human activity, we come face to face with numerous causes of stress and the symptoms of anxiety and depression.

"Nothing gives one person so much advantage over another as to remain always cool and unruffled under all circumstances." -Thomas Jefferson

THE DYNAMICS OF STRESS

In a challenging situation the brain prepares the body for defensive action-the fight or flight response by releasing stress hormones, namely, cortisone and adrenaline. These hormones raise the blood pressure and the body prepares to react to the situation. With a concrete defensive action (fight response) the stress hormones in the blood get used up, entailing reduced stress effects and symptoms of anxiety.

When we fail to counter a stress situation (flight response) the hormones and chemicals remain unreleased in the blood stream for a long period of time. It results in stress related physical symptoms such as tense muscles, unfocused anxiety, dizziness and rapid heartbeats. We all encounter various stressors (causes of stress) in everyday life, which can accumulate, if not released. Subsequently, it compels the mind and body to be in an almost constant alarm-state in preparation to fight or flee. This state of accumulated stress can increase the risk of both acute and chronic psychosomatic illnesses and weaken the immune system.

Stress can cause headaches, irritable bowel syndrome, eating disorder, allergies, insomnia, backaches, frequent cold and fatigue to diseases such as hypertension, asthma, diabetes, heart ailments and even cancer. In fact, Sanjay Chugh, a leading Indian psychologist, says that 70 per cent to 90 per cent of adults visit primary care physicians for stress-related problems. Scary enough. But where do we err?

Just about everybody-men, women, children and even fetuses-suffer from stress. Relationship demands, chronic health problems, pressure at workplaces, traffic snarls, and meeting deadlines, growing-up tensions or a sudden bearish trend in the bourse can trigger stress conditions. People react to it in their own ways. In some people, stress-induced adverse feelings and anxieties tend to persist and intensify. Learning to understand and manage stress can prevent the counter effects of stress.

Methods of coping with stress are aplenty. The most significant or sensible way out is a change in lifestyle. Relaxation techniques such as meditation, physical exercises, listening to soothing music, deep breathing, various natural and alternative methods, personal growth techniques, visualization and massage are some of the most effective of the known non-invasive stress busters.

STRESS CAN BE POSITIVE

The words 'positive' and 'stress' may not often go together. But, there are innumerable instances of athletes rising to the challenge of stress and achieving the unachievable, scientists stressing themselves out over a point to bring into light the most unthinkable secrets of the phenomenal world, and likewise a painter, a composer or a writer producing the best paintings, the most lilting of tunes or the most appealing piece of writing by pushing themselves to the limit. Psychologists second the opinion that some 'stress' situations can actually boost our inner potential and can be creatively helpful. Sudha Chandran, an Indian danseus, lost both of her legs in an accident. But, the physical and social inadequacies gave her more impetus to carry on with her dance performances with the help of prosthetic legs rather than deter her spirits.

Experts tell us that stress, in moderate doses, are necessary in our life. Stress responses are one of our body's best defense systems against outer and inner dangers. In a risky situation (in case of accidents or a sudden attack on life et al), body releases stress hormones that instantly make us more alert and our senses become more focused. The body is also prepared to act with increased strength and speed in a pressure situation. It is supposed to keep us sharp and ready for action.

Research suggests that stress can actually increase our performance. Instead of wilting under stress, one can use it as an impetus to achieve success. Stress can stimulate one's faculties to delve deep into and discover one's true potential. Under stress the brain is emotionally and biochemically stimulated to sharpen its performance.

A working class mother in down town California, Erin Brokovich, accomplished an extraordinary feat in the 1990s when she took up a challenge against the giant industrial house Pacific Gas & Electric. The unit was polluting the drinking water of the area with chromium effluents. Once into it, Brockovich had to work under tremendous stress taking on the bigwigs of the society. By her own account, she had to study as many as 120 research articles to find if chromium 6 was carcinogenic. Going from door to door, Erin signed up over 600 plaintiffs, and with attorney Ed Masry went on to receive the largest court settlement, for the town people, ever paid in a direct action lawsuit in the U.S. history-$333 million. It is an example of an ordinary individual triumphing over insurmountable odds under pressure. If handled positively stress can induce people to discover their inherent talents.

Stress is, perhaps, necessary to occasionally clear cobwebs from our thinking. If approached positively, stress can help us evolve as a person by letting go of unwanted thoughts and principle in our life. Very often, at various crossroads of life, stress may remind you of the transitory nature of your experiences, and may prod you to look for the true happiness of life.

STRESS THROUGHOUT EVOLUTION

Stress has existed throughout the evolution. About 4 billion years ago, violent collision of rock and ice along with dust and gas, led to the formation of a new planet. The planet survives more than 100 million years of meltdown to give birth to microscopic life. These first organisms endured the harshest of conditions-lack of oxygen, exposure to sun's UV rays and other inhospitable elements, to hang on to their dear life. Roughly 300,000 years ago, the Neanderthals learnt to use fire in a controlled way, to survive the Glacial Age. And around

30,000 years, Homo sapiens with their dominant gene constitutions and better coping skills, won the game of survival. Each step of evolution a test of survival, and survival, a matter of coping with the stress of changing conditions.

Millions of trials and errors in the life process have brought men to this stage. Coping with events to survive has led men to invent extraordinary technologies, beginning with a piece of sharpened stone. From the viewpoint of microevolution, stress induction of transpositions is a powerful factor, generating new genetic variations in populations under stressful environmental conditions. Passing through a 'bottleneck', a population can rapidly and significantly alters its population norm and become the founder of new, evolved forms.

Gene transposition through Transposable Elements (TE)- 'jumping genes', is a major source of genetic change, including the creation of novel genes, the alteration of gene expression in development, and the genesis of major genomic rearrangements. In a research on 'the significance of responses of the genome to challenges,' the Nobel Prize winning scientist Barbara McClintock, characterized these genetic phenomena as 'genomic shock'. This occurs due to recombinational events between TE insertions (high and low insertion polymorphism) and host genome. But, as a rule TEs remain immobilized until some stress factor (temperature, irradiation, DNA damage, the introduction of foreign chromatin, viruses, etc.) activates their elements.

The moral remains that we can work a stress condition to our advantage or protect ourselves from its untoward follow-throughs subject to how we handle a stress situation. The choice is between becoming a slave to the stressful situations of life and using them to our advantage.

WOMEN AND STRESS

Women, in these days, have a lot of balancing to do between home and workplace, including balancing between social and personal requirements. The issues of maternity, menopause, parenthood, gender roles, conditions at home and workplace, familial and social support et al, often blight women's lives in the long run.

Marital Conflict

The traditional marriage vows are emotionally moving and express a noble commitment: "I take thee, for better or for worse...until death do us part." However, we often come to dislike many things about our partner, leading to serious conflicts. Indeed, although all start with sincere intentions, almost 50% of all marriages end in divorce, in spite of enormous pressures to stay married. Why the pressures? If marriage is considered a sacred public pledge or even "a union made in heaven," then divorce might be regarded a sin (like in the Catholic Church) or, at least, a violation of a solemn promise. In addition to external pressures from family and divorce courts, there are also intense personal needs to "make it work" because it seems as though "you have failed" if your marriage fails.

Many marriages fail but do not end in divorce-the so called "empty shell" marriage. These marriages may not have intense conflicts; indeed, they may be void of feelings. There must be disappointment in such marriages, however. Let's look at some of the sources of conflict in the traditional marriage. Most married people initially try to build a smooth, close, safe relationship, preferably one without friction. In this process, sometimes the roles for husband and wife become very rigidly defined; there is no freedom, no room for growth or change. Sometimes people think they need to pretend to be or feel some way to appeal to their spouse; there is little honesty and intimacy if you think your spouse may not accept you as you really are, i.e. for better or for worse.

Fullerton (1977), in the mid-70, explained how "the perfect wife" becomes sad and angry. A woman with self-doubts may be unusually anxious to please her new husband. She tries to do everything the way he would want it done. She believes: "if I'm the good, perfect wife, I will be loved." Eventually being perfect with housecleaning and diapers and children gets tiresome and boring. She becomes resentful. Some evening when her husband arrives home from work late and finds her still mopping the floor, he asks, "Are you still cleaning?" She bursts into tears. She cries because the only ways she can vent her frustration are either to go into a rage against her husband (which she-the perfect wife-can't do) or turn her anger inward on herself. Her self-criticism increases, she clings more desperately to the husband, and feels more and more like crying.

The 1970's "perfect wife" was also prone to be jealous. According to Fullerton, a female was likely to get her sense of worth from a male-her father, her boyfriend, her husband, and later her sons. She may have gone from being Daddy's little girl to being someone's wife without ever becoming a person. She was dependent on her looks and on being a "good girl" and "perfect wife" in order to be loved. She saw her husband as having strength and purpose; he was her whole life. Even when he was at work, she carried on an inner dialogue with him. She made her decisions in terms of what he would want and expect being so needy and unsure of her worth.

.Abused women are 1.5 to 3.0 times as likely as women never abused to abuse) pain, dizziness, gynecological conditions, abortions, and have a health problem. These long-term difficulties may be (beyond injuries) from mental problems, Women living in a rich including suicide. So the connection between stress and health is seen again. Country that had been abused was less likely to still be being abused. Apparently, when they can escape, a percentage does get out.

How common is wife rape? Several studies have found that about 10% to 15% of women who have been married report being sexually to have sex with someone is assaulted by husbands or ex-husbands (Rape Center). These figures may be low because women are less likely to consider sexual assaults by husbands as "rape." Laws against rape in marriage have only been passed in recent decades. That is strange; because the use of force and threats wrong, it doesn't matter if the woman is pulled into an alley and raped by a total stranger or if an angry husband demand scoring. It is all rape. Wife rape is reported by 50% ds sex now or if a date insists on of women in shelters in America, 1992.

Abuse within the family

About 10% of all violent crime is family violence (committed by someone within the family). Much abuse is still hidden, not only is marital abuse kept a secret but sibling abuse is also. Within the privacy of our homes and even unknown to the parents, brothers and sisters physically, emotionally, and sexually mistreat each other (Wiehe, 1990). Some good news is that family violence has declined in the last 10-15 years along with an overall reduction in violent crimes.

Recently several studies have looked at the long-range consequences of abuse or "adverse childhood experiences." This includes a wide variety of hurtful, stressful events for children-actually if a child has suffered one abusive experience, that child has an 80% increased risk of being abused in some other way as well. There are more of these bad experiences than most of us realize, e.g. more than half of middle-class children enrolled in the Kaiser Permanente plan has had one such experience, one in four children have had two types of abusive experiences, one in 16 have experienced four types of harm. Then when abused children are followed over 50 years or so, a remarkable array of health, psychological, and behavioral consequences are found (more in formerly abused children than other children). Examples: they smoke more, have more depression, experienced more anger, are more abused by partners, have attempted suicide more often, used more illegal drugs, have heart disease, diabetes, obesity, alcoholism, and do more poorly in their job (Sawires, 2003). There are several other long-term studies that confirm these findings. Even though we are forewarned, our physical and psychological professionals are not good at prevention.

Weaknesses as battering men and are in a situation where they can physically abuse their partner. In the 1990's it was known that women are victims of 11 times more reported abuse than men (Ingrassia & Beck, 1994). That may be changing. But, as mentioned, men may be hesitant to label themselves as "battered husbands." Spouse abuse occurs in all social classes and with independent as well as dependent women and men. Society, relatives, strangers, neighbors, and the police don't know how to deal with family fights but society pays the bills in the emergency rooms, in marriage counseling, and in divorce court.

Abuse should not happen but no treatment is a sure cure; in fact we don't even have a good cure. About half of male batterers will not get treatment and half of those that do, drop out. Little has been written about treatment for female batterers. In most cases, it is wise to report male batterers and their abuse to the police. Most police have had some training in handling "domestic violence" cases; however, officers in New York, which has a mandatory-arrest law, arrest only 7% of the cases and only report 30% of the domestic violence calls (Ingrassia & Beck, 1994). Police are supposed to provide the victim some protection (of course, this is hard to do and can't be guaranteed). Recent research confirms the benefits of pressing charges in abuse cases. If the abuse is not reported to the police, about 40% of the victims were attacked again within six months. If the abuse is reported by battered wives, only 15% were assaulted again during the next six months.

So, protect yourself.

Almost no one asks the question should women who batter their husbands be reported to the police? Well, it seems fair that women batterers should be reported to police like men are...but women do not kill partners at the same rate as men do. And there are other buts: what would this reporting do to the already shaky relationship? And how would Police handle that task? Police have enough trouble taking male batterers seriously; do we know how they would deal with female batterers?

Why do Women Stay

To the outsider the real question is: Why do they stay together? Why doesn't she leave? Or, why doesn't he leave? Why should she have to leave instead of him? If they stay together, there must be varied and complex dynamics which tie an abusive couple together. We have much speculation; we need more facts. Clearly, there are likely to be emotional bonds, fears, shame, guilt, children to care for, money problems, and hope that things will get better. Many abused women are isolated and feel unable to find love again. Some women assume abuse is their lot as a woman; this is-for them-an expected part of life. A few women even believe a real, emotional, exciting macho "man" just naturally does violent things and feels superior to women. Some violent men are contrite later and even charmingly seductive. Some women believe they are responsible for his mental turmoil and/or are afraid he will kill himself or them. She may think she deserves the abuse. Many believe he will beat them more or kill them, if they report the assaults to the police. Of course, injury and death do occur. The abused woman often becomes terrorized and exhausted, feeling totally helpless. Walker (1979, 1993) says the learned helplessness (within a cycle of violence and making

Indian Widowhood - The Silent Cry that Society Must Hear

India is a country of holy cities like Varanasi, Tirupati, Vrindavan and a country with thousands of monks and thousands of female deities and yet, unprecedented socio-cultural discrimination against widows persist. The same society denigrates a woman while puffing her on a pedestal inside a temple. What is the choice that women have? In the words of Prime Minister Atal Behari Vajpayee "Our society does not always treat widows, especially young widows well" and he goes on to emphasize that "Social and religious establishments can play a major role and society at large needs to be sensitized so that widows are treated with dignity within their families".

It is indeed high time that after 50 years of independence, we try and understand the factors contributing towards discrimination of widows and to identify measures for their rehabilitation. We have to work hard to generate public opinion against the practice of ostracism and to search for ways and means to empower the girl-child.

Today there is an urgent need to bring about attitudinal changes in society towards widows, their rights to assure empowerment and implementation of their legal rights. Due to male patriarchy and vicious customs even the most worthless of men have been enjoying superiority over women!

Changes in society from time to time, have influenced the status of widows and the behaviour of society towards them. Despite their social, religious and regional differences, women had to readjust their entire lives after the demise of their husbands. In our country there are four times as many widowed females as widowed males.

Today there are 33 million widows in India 10% of the female population. Every fourth household in India has a widow. 50% of the widows are over 50 years. A simple study of 1994 shows that of the 88% widows who remained in the deceased husband's village after his death, only 3% shared the same hearth with her in-laws. Also, less than 3% widows lived with their parents. In the age group of over 60 years 64% are widows. Widow pensions range from Rs.100 to Rs.500. On a study of seven states of rural India there were 28% widows eligible for pensions only 11% were actually receiving it. There are approximately 20,000 war widows in India from the '62,'65,'71 and '99 aggressions.

The new millennium calls upon us to get into action against this social callousness, otherwise we will be found guilty against our mother and grand mother. This national ignominy is our collective responsibility.

Are we not really responsible for the depersonalization of the widows? Are not the parents guilty by perpetually referring to her as *'Paraya Dhan'* (property belonging to other) thus sowing the seed of insecurity in her mind. Without the support of her own family she is an easy victim of brutalities practiced by her in-laws or the society. Are not the in-laws guilty, because they do not wish to give her rightful share in the family property? What about the religious leaders, who justify these sanctions against a widow on religious grounds and do not openly, speak against them and the State, which will not enact or enforce laws that guarantee single women the right to an honourable existence. Media which consistently portrays her as a suppliant creature, born and raised to unquestioningly serve the men in her family showing her as a helpless victim of neither male violation who neither protest nor rebel.

Today unfortunately our political leadership also turns a blind eye to their plight because they do not form a vote-bank, and are fearful of raising a controversy. Last but not the least we as women also are the main protagonists and silent witnesses of atrocities practiced on widows. In the words of Human Resource Development Minister Shri Murli Manohar Joshi "The status of widows in our society is not, what it should be. They are often subjected to harsh treatment by the family and society. Widows are very vulnerable and deprived of access to resources and in some states widows are either driven out of the home or abandoned in holy cities like Vrindavan or Varanasi where they live a life without dignity and honour".

In the holy city of Vrindavan and surrounding areas there are more than 10,000 widows. The women are mostly widows, but some who have been abandoned by husbands, are brought by relatives to Mathura, Vrindavan and Kashi on the pretext of being taken on a pilgrimage. Once there, they are left in a strange and unknown city to fend for themselves. The younger ones are soon pursued by unscrupulous elements and forced into prostitution. Their bodies are bartered like commodities. These helpless women have no one to turn to. The Dharamsalas (transit homes) do not allow extended stay and with rising prices shelter has become unaffordable. Instead of the *"Moksha"* (salvation) they were promised, there is destitution and humiliation. Their shrouded figures with blank looks and shriv-elled bodies, pace the lanes and by-lanes of Mathura, Vrindavan and Varanasi. They are the disowned, the dispossessed-forced to live on the fringes of Society. For most of them begging is the only source of income.

Those who perpetrate violence on women are not necessarily anti-social elements, but their own kith and kin. They have deliberately turned their backs on them. Should society do the same? Do they not deserve a helping hand to pull them out of this mire into which they have been pushed? Till such time this is done the entire Indian society stands indicted.

The constitution of India says that all men and women are equal; however, the society has made women and widows into bonded slaves. Today the vital need is to sensitize the society and the mindset which considers widows harbingers of misfortunes.

It is a sad reflection on society that religion is being used as a tool to oppress the widows and inflict all sorts of atrocities on them. His Holiness Jagatguru Shankaracharya (of Joshimath) was of the opinion that widows are like sadhus to the world and that is why they wear white clothes and have their hair shorn, so that society recognizes them as an ascetic and since they have been given a role of an ascetic they are not invited on happy occasions. They are supposed to be dedicated to God. He ruled out widow remarriage on the grounds that it was against the Shastras.

An amendment is needed for the Hindu Succession Act and Muslim Law, where a man can disinherit his wife/widow. These reforms have been introduced in advanced countries like Canada. All over the world the women have the right of inheritance and it is time that we also do the same in India. It has been proposed in the Christian law, all property passes on to the widow and the sons and daughters till she remarries. However, if the male member makes a will, he can legally disinherit his widow. So reform against this law should be introduced.

It is ironical that during the Rig Vedic times women had the right to property including the *Streedhan.* Today, the customary law contradicts. With the support of amended laws there can be major changes that can benefit women. This should include property rights as well as *streedhan,* inheritance from ancestral property, selfacquired property from husband and also the right to maintenance. 90% of women, who are impoverished, destitute and cast away with children, do not even get maintenance from their husbands. How can they fight for their rights? A civil servant, educated and sophisticated, violated the law by manipulating that his wife was mentally deranged and produced one psychiatric report and the prescription for the treatment. Is it not strange that women are de-barred from inheritance, as only a chaste and a sane woman is allowed under the law to inherit her husband's property?

What Hindu Texts Say

"In Yajurved (Taithriya Aranyaka-6/14), one reads as follows "Get up, oh woman, you who lie down by the side of this your lifeless husband. Come to this crowd of living people about you here and may you become the wife of some person desirous of taking the hand of a widow in remarriage". "In the later periods, Kautilya *Arthshastra* (1, 12, 4.5) mentions that a poor widow of high family, clever and anxious to earn her livelihood can become a woman priest Parashar and Narad Samhitas mention that in the event of husbands death or impotency or his leading a life of an ascetic or his getting can remarry." and today we find that there are vested interest in running the Bhajanashrams in these holy cities. Many of these widows who had a large bank balance have died and large amount of money is left in the banks which need to be taken out. Some of the measures that we can take to improve the situation of widows are

Action Plan:

1. Create awareness in the public, the media and administration regarding the plight of widows through workshops and seminars thereby providing a forum of interaction for NGO's engaged in the field of rehabilitation of widows and the administration in the different states.
2. Persuade religious leaders to openly condemn the brutalities perpetuated on widows, so that they may live their lives with dignity and honor.
3. Encourage NGOs and Charitable Trusts to provide shelters for widows and abandoned women, particularly in religious places like Varanasi, Mathura, Rishikesh, etc. These shelters should also run rehabilitation programs so those women can become productive members of society.
4. Persuade government agencies to enforce existing laws and guidelines regarding social security of widows. Pension entitlements should be disbursed speedily without creating

undue bureaucratic hurdles and transfer of property rights expedited.

5. Ensure that inheritance rights, as governed by law, are enforced by *Panchayats*. It has been the experience of agencies working in this field that customary law prevails over statutory law particularly in rural areas.
6. Convince the Law Ministry to amend the Hindu Succession Act 1956, by revoking the provision according to which a man may will away his entire property and assets.
7. Enable widows and single women to access education, counseling and training in order to attain economic independence, by reserving seats for them in educational training institutions. With this in mind sometime ago submitted a Charter of Demands to the Government is as follows:

A. The National Policy of Empowerment of Women be passed as early as possible

B. Recommendations of the National Commission for Women are implemented without further delay.

C. A National Policy on Health with special reference to marginalised groups is passed.

D. Widows comprise ten per cent of total women's population, hence

Their Legal Rights be Ensured:

- Their right to property be ensured through joint ownership of land pattas/ agricultural and housing.
- Registration of marriages be made compulsory.
- Pensions for widows and deserted women be revised to provide for the present day needs and be made compulsory.
- Support services and rehabilitation measures like shelter, employment/training measures for widows and deserted women be given priority in all government programs.
- Special programs for income-generation be devised for widows.
- Children of widows belonging to weaker sections be given preference in school and college education.

To Religious Leaders:

- Interpret religious text and traditions in the right spirit.
- Discourage undermining of women's prestige and social norms which ostracize and marginalise women.
- Gear religious discourses to change attitude towards women.

To Civil Society and Public:

- Public should support efforts made by Government of India.
- Each family to be responsible in restoring self-esteem of widows and deserted women and assist in building a climate of moral concern and community care for rehabilitation of widows and deserted women.
- Monitor such families which indulge in violence against helpless women.
- To Educationists and Educational Institutions.
- Impart Human Rights Education and prepare text books to be gender friendly.
- Care of elderly be part of a child's education.
- Give more emphasis on skill development.

To Media:

- Project success stories of brave women.
- Educate public through slogans and hoardings. Government's schemes to be disseminated through mass media.

To NGOs:

- Encourage community based organizations to mobilize people's participation towards upliftment of widows and deserted women.
- Facilitate networking and coordination among like-minded individuals/organizations working for welfare and rehabilitation of widows.
- Extend help to widows and other marginalised women to solve their problems by counseling.
- No society can be free unless we provide social security

measures which include education and health. We have to talk about pro-women measures and rehabilitative measures.

- The flow of water from Ganga and Yamuna has seen thousands and thousands of widows silently cry to take action against atrocities perpetuated on them.

SOCIAL AND WORK STRESS

Sociological researches assert that family structure (working or stay-at-home mothers and other models) affects performance and employee attendance either directly or interactively. Family demands and family attitudes were found to influence the absence frequency at workplaces.

Experiencing a high level of burnout was associated with increased absenteeism if employees had children under six living at home or reported having difficulty with their child care arrangements. There is a strong relationship between social support and mental stress and trauma in women. It relates to a woman's help-seeking attitude, social networks, kinship networks, and support networks. Besides these factors, adapting to a new workplace culture and reformations tends in job sectors, affects men and women alike.

GENDER ROLE

Quantity and quality of leisure time distribution between the genders is an interesting index of how women get burdened with stress for either natural or social obligations. Multinational Time Budget Data Archive and the Australian Time Use Survey suggest that women are now bearing a "dual burden" as both family providers and family careers. Although men and women have similar quantities of free time, when the character of leisure is considered the gap between genders reemerges. Absence of reciprocal and joint emotion management within family is a nagging stressor for women. Mostly mothers handle the bulk of the parental responsibility such as educational and emotional care of children. This can be physically both and psychologically draining.

MENOPAUSE

Beside PMS, the other most nagging stressor coming from hormonal changes in women's body is the menopausal phase in one's life. Women go through a lot of anxiety, tension, worry, emotional suppression, and a lot of physical and mental stress, before and during menopause, even if they are aware of its eventuality.

Doctors define menopause as the time after 12 months have passed since your last period. Every woman experiences menopause differently, starting from the age of 30 to 50s. Signs and symptoms are also very individual. One may breeze through menopause with few symptoms, or may experience a number of physical and emotional changes. Changes such as:

- Unpredictability of periods.
- Realization of uncomfortable truths such as loss of fertility and sexual capability.
- Changes in appearance due to fat accumulation in body and emergence of facial hairs (presence of small amount of the male hormone testosterone).
- Hot flashes occurring in the upper body (the face might look flushed, and red blotches may appear on your chest, neck and arms lasting from 30 seconds to as long as 30 days, one may feel chilled, weak and slightly faint).
- Sleep disturbances and night sweats followed by chills (one in four midlife women experience insomnia. Lack of sleep may affect one's mood and overall health), ultimately leads to the predictable result of stress and emotional strain in women.
- As a women goes through menopause, women may experience mood swings, be more irritable or be more prone to emotional upset.

  Other stress factors that can occur in this stage of adulthood such as the illness or death of a parent, grown children leaving home and in some cases, change in the pattern of emotional relationships with one's spouse, add to the ongoing physical stress. To minimize and manage the psycho-physiological effects

- Schedule regular checkups and consultations with your physician about diet, nutrition, drugs or hormone administration.
- Be open to adopt a positive lifestyle change, which could be physically and spiritually fulfilling and suitable for your personality type.
- Avoid caffeinated beverages and exercise right before bed to avoid sleep trouble and insomnia. Practicing relaxation techniques such as deep breathing, guided imagery and progressive muscle relaxation does help.
- Apart from identifying what helps trigger hot flashes (smoking, spicy food, hot beverages, alcohol, warm room et al), getting regular exercise and dressing in layers can manage the condition.
- Wear cool cotton clothing to bed and keep an extra set handy, in case of night sweating.
- Taking a balanced and nutritional diet is a must. During this phase body needs at least 1,200 to 1,500 mg of calcium supplements.
- Abstain from smoking. Smoking can increase the risk of heart diseases, stroke, cancer and a range of other health problems.
- Take stairs instead of an elevator or park farther away from destination. Better still walk down to your destination to stay positively active. Vigorous exercises such as aerobic, brisk walking, jogging and dancing, and strength-training exercises, can be excellent protection against cardiovascular disease, diabetes and osteoporosis, and most importantly stress.

MEN

Much of the causes of male depression and stress arise from their self nurtured identities, especially related to their professional status. "If you ask a man who he is, the first thing he says is his work- I'm an executive, I'm a doctor, I'm a house builder," says Glenn E. Good, an associate professor of educational and counseling psychology at the University of Missouri, Columbia. "Suppressing feelings and internalizing stress are acquired male traits", says Good, "On some inner level, it comes down to: If I can't tough it out, then I'm not much of man."

WORK STRESS

For men workplace stress can have extreme consequences. In Japan, work stress related suicide rate among men has risen over the last 15 years. According to the Government's Statistics Bureau, the highest suicide rate occurs in men from 35 to 44 years old, making it the 13th most common cause of death for men. Newspapers carried the triple suicide story in March 1998, when three Japanese men-all heads of car part companies-took their lives on the same night. The reason they gave is- Poor Company Finances!

Uncertainty in workplace can cause high levels of stress. Lack of information or the actions of other people can negatively affect your ability to perform. Causes of uncertainty can be:

- Not having a clear idea of what the future holds
- Not knowing where your organization will be going
- Not having any career development plans
- Not knowing what will be wanted from you in the future
- Not knowing what your boss or colleagues think of your abilities
- Receiving vague or inconsistent instructions

WHAT CAN HELP

- Learning to communicate-even if it means transcending traditional male egos. Trying to openly resolve conflicts at work and even seeking outside support from counselors, support groups or co-workers can bring a sea change in your attitude to handle stress. "When men start admitting their insecurity," says Hope Hills, president of Circle Consulting Group in Milwaukee, "especially to their colleagues, there can be a real change in their level of stress and comfort with themselves."

- Learning not to give much importance to each and every work that comes in your way.
- Making the body stronger and more able to withstand stress is an important step towards managing stress. Learning relaxation techniques-such as meditation, biofeedback, deep breathing or yoga-for 15 to 20 minutes during the work day, can be effective stress busters:
- Learning to recognize a stressor.
- Not putting unjustified demands on yourself to perform better and more.
- The most effective way of countering cases of uncertainty in workplace is to ask for the information you need. If you ask in a positive way, generally, people are usually quite happy to help.

MIDLIFE CRISIS

This syndrome is often known as the male menopause and occurs between the ages of 35-50. The phase is often described as the end of a beginning, and beginning of an end. It is much like adolescence revisiting you once again, but with a different agenda.

This transitional point emerges with two fundamental elements of crisis where there is a sense of urgency along with intense emotions and confusion. According to a description of midlife crisis in the lesson4living (lifestyle website), "At midlife the meaning and purpose of the first half of life begins to fail. The agenda is no longer focused on differentiation but integration. We must give up the identity we worked so hard to get so we can become more."

MYTH OR FACT

Though most men deny the existence of any such phenomenon as 'midlife crisis', there are tell-tale signs of behavioral and attitudinal changes on their part. In the early stages-at around 35-men tend to avoid the topic saying, "I'm too young to have it". During the typical age-range-40 to 50-they become defensive with explanations such as, "It is not happening to me". While on the finishing line they may try to avoid the topic dismissing its existence altogether.

Warren Farrell, author of Why Men Are The Way They Are, has denounced the male midlife crisis as a myth. He maintains that the transition holds true for both men and women. For women this phase occurs at the age when the last child leaves home, which generally concures with the onset of menopause. For men, the phase comes when they feel that they have reached a quiescent period in their career. Farell feels that the phase is not a 'crisis' as it is thought to be. He explains that since most men are at the peak of their career and influence around this period, they see this time an opportunity to achieve and do all that they couldn't do earlier, however capricious the methods may be.

But, the facts show that 'crisis' or 'not' midlife in men is definitely metamorphic in character.

SYMPTOMS OF STRESS

According to the Briggs-Myers model (First, identified by the psychologist Carl Jung in the early part of the 20th century, and expanded on by Katherine Briggs and her daughter Isabel Briggs-Myers) one may show such symptoms of 'Mid-life transition/crisis', as:

- Discontentment with life and/or the lifestyle that may have provided happiness for many years.

  ("Midlife is when you reach the top of the ladder and finds that it was against the wrong wall." -Joseph Campbell. The ladder can be career or personality growth, marriage or family.)
- Boredom with things/people that have hitherto held great interest and dominated your life. (After 20 years of married life you may suddenly wake up to the feeling of being trapped in the relationship. There could

be a confusion-of feeling burdened, and yet oddly free at the same time. It is possible that you are avoiding a confrontation with midlife transitions, and your own disappointments are being projected on to the marriage. The feeling can make you mentally disconnected with your spouse)

- Feeling adventurous and wanting to do something completely different. (You may be tempted to do things that are uncharacteristic of your age or maturity-frequenting discos; showing interest in extramarital relationships-almost always with women half your age or drawn towards alcoholism and substance abuse. Even though our preferences are innate, our behaviour and perception of ourselves are modified in order to 'fit in' with the various situations (personal/social) in our life. Jung called this process 'Accommodation'. Especially, such 'accommodated' individuals tend to break free from their set 'personae' during this transition period.)
- Questioning the meaning of life, and the validity of decisions clearly and easily made years before.
- Confusion about who you are, or where your life is going.

TAKING IT HEAD-ON

All transitions are difficult, but, it is better to take the changes as something normal, natural and unavoidable. It helps to take the journey by equipping yourself with a roadmap-understanding the point where you are and what your destination should be.

You are likely to face another major identity crisis. The demand to depart from your established identity is greatly resisted from inside. You must let go much of the experiences that piggyback on you into midlife.

It often prompts you to look back in life and search for a better meaning of it. It prompts you to ask, "Is this all that there is?" you must be open to this awakening, accept the challenge to investigate life more. It can be a facilitating process in becoming a 'whole' person, through personal growth.

- Many times people tend to retreat into the pattern of clinging on to youth and try in vain to recapture it. The anxiety, frustration and depression born out of it can lead one to a blind alley. As life means progress and not stagnation, the golden rule could be learning to move on with it, and avoid stress by remodeling one's thinking and attitudes towards life.
- How to cope with your husband's midlife crisis: Have patience. Let him learn to cope with the new feelings and emotions that he is encountering. Don't try to help him understand his problem, as the social conditioning of behaving 'like a man' may make him feel insecure and put him on the defensive.
- Give him space. If he wants to spend less time with family and more time with his friends, don't cling, berate or belittle him. His behavior may make you feel insecure and even cause anger in you, but, retaliation on your part can make things worse. Even if he blames you for his disenchantment with the present life, don't relent. Don't take the responsibility for his midlife turmoil. Let him identify the real cause of his problem. There is no point suggesting him to go for counseling since he is unlikely to accept that he has a problem. It might help, if you rather try to cope with your feelings and the overall situation.
- If he is feeling insecure, let him know that you love him and are with him. But, make it a point not to expect any thing in return, even though it is difficult to practice. Even though you played no part in the creation of his crisis, how you behave at this juncture will define your future relationship.

HOMOSEXUALITY: THE BASICS

Sexuality is a life long process about learning to decide what is right for you, and learning to respect what others feel is right for them. This is why there is no right or wrong about the kind of partner you might desire. There are many women in this country who are attracted, and live with women, and there are many men who do the same with other men. The reason these people can not be open about their relationships is because they feel that if they are, the society they live in (which includes their friends and parents, colleagues etc.) would be rejecting and judgmental. This section hopes to address your myths and problems.

Am I Gay

Many people have feelings towards other people of the same sex, and wonder whether this means that they are gay. For many people these feelings can be very intense and alienating. Some people who are attracted to other people of the same sex are gay and go on to have sexual relationships with people of the same sex. But other people who have gay feelings find that these change over time and they become attracted to people of the opposite sex. Other people are attracted to both men and women, and have relationships with both. Some people are not attracted to anyone and wonder if this is a sign of homosexuality. Often it is only time that will resolve these feelings. If you think you might be gay and feel you need to talk to someone most countries have telephone helplines and organisations that can provide information and support for you.

When do People Know that they are Gay

There is no simple answer or standard answer to this question, as it vari from person to person. Generally it can be said that being gay is not something a person suddenly begins to consider, and it may not be something they can initially put a name to. Research published in 1996 showed how the young gay men interviewed had described a set of feelings which they gradually realised made them 'different' in some way, and a set of feelings they thought maybe every teenage boy has.

'I thought, well, this is just the phase bit. Sooner or later I'm going to start finding women attractive. I never did. As I became more attracted to men, and I still wasn't getting attracted to women, I thought, shit, I'm gay. And it was really quite a shock when it hit me.'

Eventually all people who are gay realize that not only are they sexually attracted to members of the same sex, but that this attraction is not transitional. This realization could come at any time during their lives.

What causes People to be Gay

'One of the things I can remember thinking a lot about is ... why am I like this? Is it someone's fault?'

There is no simple answer to the question, 'Are some people born lesbian or gay?' There are some theories which stress biological differences between heterosexual and homosexual adults, suggesting that people are born with their sexuality already determined.

In 1993 the American researcher Dean Hamer published research that seemed to prove that homosexual orientation could be genetically transmitted to men on the x chromosome, which they get from their mothers. However when this study was duplicated it did not produce the same results A follow-up study which Hamer collaborated on also failed to reinforce his earlier results. Most recently research published in April 1999 by George Rice and George Ebers of the Universty of Western Ontario has cast doubt on Hamer's theory. Rice and Ebers' research also tested the same region of the x chromosome in a larger sample of gay men, but failed to find the same 'marker' that Hamer's research had produced. Claims that the part of the brain known as the hypothalamus is influential in determining sexual orientation, have yet to be substantiated. At the moment it is generally thought that biological explanations of sexuality are insufficient to explain the diversity of human sexuality.

'How can science tell you what I am? I mean I've had boyfriends, and was happy with them, had girlfriends and may have boyfriends again for all I know. If it's a gay gene what's going on? Is it just turning itself on and off in my head? It doesn't feel like biology it feels like love.'

Psychosocial explanations offer a variety of factors that could contribute to the development of a person's homosexuality. For example, a female dominated upbringing in a gay man's past, with an absence of a male role model. Others stress adherence or deviance from conformity to gender roles, and individual psychological makeup. While none of these factors alone completely answers the question 'what causes homosexuality?', they rule out some things. For example, lesbian and gay young people are not 'failed' heterosexuals. Also, homosexual partners are generally of the same age proving wrong the assumptions that young people are 'turned gay' by older people.

What is clear is that people's behaviour is influenced by their family environment, their experiences and their sense of themselves. Beliefs about sex are initially shaped by family values. Later on these beliefs may be shaped by pleasant and unpleasant experiences of sex and also shape their choice of activities and partners. Throughout their life a person's sense of who and what they are has a strong impact on their sexual development and experience.

SOUTH ASIAN DISABLED WOMEN AND MULTIPLE OPPRESSION

This is an attempts a look at some of the different forms of oppression faced by disabled women in South Asia. Among the perceived causes of oppression, apart from the two key values related to gender and disability, the paper touches the impact of ethnic, caste and religion differences, rural versus urban differences. The paper proposes that some of the cultural and social values related to gender roles and hierarchies in family, influencing or determining different forms of oppression affecting South Asian disabled women may be common to the region, over-riding ethnic, caste and religion differences and geographic lines. Finally, the paper also briefly discusses the relevance of social model of disability to South Asian Disabled persons.

Understanding Oppression in Relation to Disability

Young has defined oppression as a "structural concept that refers to the immobilisation or diminishing of a particular social group." (Young I., 1990, cited Priestley M., p. 86, 1998).

She explains five features, which characterise social oppression – exploitation, marginalisation, powerlessness, cultural imperialism and violence (Young I., 1988, cited Priestley M., p. 86, 1998).

'The oppressors overwhelm the oppressed with their values and norms, which effectively silences people. In this context Freire uses the concept of "myth". By pressure from those in power, the oppressed have internalised those myths, which we can speak of here as "lies" because they have been purposefully and knowingly imposed upon the people without taking into consideration their reality. The oppressed people feel ignorant and they become dependent on the culture of the oppressors, the so-called "experts", specialists in society. The needs of the oppressed and the knowledge gained from their own "experience" are not regarded as important; they are ignored, devalued and considered as inferior. (Fritze, 2000)

Thus minority population groups perceived as being "different" from the majority populations and because of their relative powerlessness, become easy targets for oppression. Some of the oppressed groups, like women and blacks, may not necessarily be minority groups. Discriminating social, cultural, political and historical traditions and systems, which help in keeping the oppressed persons fragmented and isolated in small groups, play a key role in the maintenance of power-relationships linked to oppression.

By their failure to conform to 'normality', they are marked out members of a 'minority group' in a similar position to other oppressed groups such as black people or 'homosexuals', because like them they are viewed as 'abnormal' and 'different' (Hunt P., p. 152, 1966)

The opportunities for disabled persons to come together and to share their similar life-stories of individual oppression and to come to a collective understanding about their social oppression, has been possible only recently, over the last century, especially over the last few decades.

'People with disabilities have been a hidden, misunderstood minority, often routinely deprived of the basic life choices that even the most disadvantaged amongst us take for granted. In the last twenty to thirty years, little noticed alongside the civil rights struggles of African-Americans, women, gays and lesbians, and other minorities, another movement has slowly taken shape to demand for disabled people the fundamental rights..' (Shapiro, J.P., 1993)

'For Abberly, oppression is an all inclusive concept, which relates to the hierarchical relations pertaining to gender, race and sexuality. Further he suggests that in the same way that feminists and antic-racist theorists have addressed biological arguments for sexism, heterosexism and racism, a social oppression theory of disability must do the same.' (Barnes & Fiorentino, 2000)

Any recognition of social injustice requires coming together of affected persons, to share information and experiences and to create a collective identity of their oppression. For the disabled persons, the technological innovations after the second world-war, especially those related to communication, made it possible to overcome their isolation and to come together for sharing of information and experiences. Over the last few decades, this has led to coming together of disabled persons, their transformation in to activists, development of social model of disability, their decision to create organisations run and managed entirely by disabled persons, their fight for equal opportunities and human rights, etc. Union of the Physically Impaired Against Segregation (UPIAS) meeting and its policy statement in 1976 have played a key role in definition of social model of disability in the UK:

'Our position is similar to that of many people who are middle aged or elderly, who have had break-downs, or are 'mentally handicapped', black, ex-prisoners, unskilled workers, etc. We are usually among the first to lose our jobs and be cast on scrap-heap when it suits the 'needs' of the economy. If we are lucky we may be drawn in again, to do the worst paid work, when business starts to boom once more. If we are unlucky, then we could face a lifetime on the degrading, means-tested poverty line. If we are very unlucky we may be consigned to a soul-destroying institution.' (UPIAS, 1976)

The disabled persons have themselves defined disability as a loss or limitation of opportunities to take part in the normal life of the community on an equal level with others due to physical and social barriers (DPI, 1982, p. 105). This social model of disability clearly contrasted with the traditional medical model, which sees disability as a result of physical or functional abnormality. Disabled persons organisations, over the last few decades, have launched movements for recognition of the role of community and social barriers in the creation and maintenance of disability. However, even in developed countries where the movement by disabled activists for equal opportunities and human rights has been stronger, there is still a long way to go to change the traditional views based on medical model of disability.

Similarly there was said to be nothing wrong with disabled people (who, like women or blacks, tended to be lumped improbably under one label...). Society must change its attitudes and construct enabling, supportive, inclusive, people-friendly environments. These worthy ideals seem, however, to have little impact on European mass perceptions. The unenlightened continue to think that if they acquired an impairment resulting in the disability of no longer being able to see their way around, or hear music, or stroll down to the pub, there would be something wrong with them, a personal problem for which society can not entirely be blamed. (Miles, 1996, p. 486)

On the other hand, in developing countries, the opportunities for disabled persons to come out of isolation and to interact with each other have been much less. There, in spite of different international and national declarations and laws, the fight for equal opportunities and social justice for majority of disabled persons is even more difficult.

Women and Oppression - Understanding Gender

The women's movement for equality and human rights has a much longer history compared to that of disabled persons. For the women's movement, it has been important to distinguish between sex (the biological difference between men and women) and gender:

Gender is a set of roles which, like costumes or masks in the theatre, communicate to other people that we are feminine or masculine. This set of particular behaviours - which embraces our appearance, dress, attitudes, personalities, work both within and outside the household, sexuality, family commitments and so on - together make up our 'gender roles'. (Mosse J.C., p. 2, 1993)

Common socially accepted gender roles for women in most of the communities see women as being inferior or weaker. In the most extreme cases, some societies promote practices like genital mutilation to keep their bodies and mind subjugated.

Multiple oppression

An individual may be oppressed for different reasons like nationality, colour of skin, gender, sexual preference, age, religion, caste, ethnic origin, disability, etc. Thus persons who are disabled and at the same time, are black or gay/lesbian can be oppressed for each of these reasons at multiple levels, separately or simultaneously.

Women are oppressed not only by their gender but also by their class, race, caste, and colour, sharing these oppressions with men from similar social groupings. But if poor men are doubly oppressed by their colour and class, poor women experience a 'triple yoke of oppression'. (Mosse J.C., 1993, p. 86)

While some of the writers tend to calculate the number of different layers of oppressions a person may be facing, some of the disabled activists have proposed that rather than being two or more separate and contained forces, the different oppressing factors are integrated and interwoven (Branfield, 2000).

'A disabled woman therefore suffers a multiple handicap. Her chances of marriage are very slight, and she is likely to be condemned to a twilight existence as a non-productive adjunct to the household of her birth. At best she may be the object of misplaced sympathy which regards her as helpless and unable to do anything for herself or for the family; or she may well be kept hidden in order not to damage the marriage prospects of siblings; alternatively, she may be turned out to beg. It is a small wonder that many disabled female babies do not survive.' (Coleridge, p. 154, 1993)

Apart from the oppression perpetuated by those with "power", long-standing oppression can also become internalised by the oppressed persons. Thus the oppressed persons may feel that they are inferior or undeserving. 'It (internalised oppression) would not exist without the real external oppression that forms the social climate in which we exist.' (Mason M., 1992)

'There is the added feeling of indebtness for all the physical assistance she receives. This feeling is so intense and ever present even when family relations are warm and spontaneous, that she would endure and suffer silently any amount of discomfort and pain rather than express her needs and requirements lest she added to their burdens, as she sees it.'

TO BE WOMEN IN SOUTH ASIA

South Asia includes countries of the Indian sub-continent – India, Pakistan, Bangladesh, Nepal, and Sri-Lanka. The region has seen powerful women like Srimaya Bhandaranaike in Sri Lanka, Indira Gandhi in India, Benazir Bhutto in Pakistan and Khalida Zia in Bangladesh, guiding the destinies of their countries. Most of them owe their roles, at least initially, to their being wives or daughters of presidents or famous men. A large number of women, especially in urban areas, are lawyers, doctors, teachers, writers, artists, social activists, etc. Even in rural areas slowly but gradually there are many examples of women assuming better control on their lives. At the same time, for a large number of women, lives in the rural and urban families, oppression is mixed in with all aspects of their lives.

'When I became pregnant for the third time, my husband took me to the doctor for scanning. He was afraid that it might be another girl. As it was the third month of pregnancy, the report was not clear whether it was a boy or a girl. I was taken for scanning during the fifth month. It was confirmed that the child was a girl. My husband didn't want a girl and he sent me to my parents' house to get the pregnancy terminated. He told me to not to come back to his house without undergoing abortion. As I could not oppose him, I got the pregnancy terminated. Before I got a baby boy, I got three pregnancies terminated. He threatened to divorce me when I spoke against. I have younger sisters, who are yet to be married. Therefore I keep quite, remaining like a puppet in my husband's hands.' (Malathi, p. 101, 1999).

CULTURE AND RELIGION INFLUENCING LIVES OF SOUTH ASIAN WOMEN

Hinduism and Islam are two principle religions in South Asia region, though significant numbers of other religions like Christian, Sikh, Buddhist, etc. are also present. Religious beliefs are closely interlinked with traditions and cultural roles of men and women in the society.

Ramayana, the mythological story of King Rama and his wife Sita, has a special religious and cultural value for the Hindu families, especially in North India and Nepal. The influence of this story on the cultures in countries like Sri Lanka, Myanmar, Thailand, Indonesia, Vietnam and Cambodia is still evident. The representations of Rama and Sita represent the ideal roles for a large number of men and women in the region:

'Although Sita's life can hardly be called a happy one, she remains the ideal woman through whom the patriarchal values may be spread far and wide, through whom women may be taught to bear all injustice silently.' (Sen N. D., 1998)

Islam is the majority religion in Pakistan and Bangladesh, though all other countries of the region have significant numbers of Muslim populations. Stories about Islamic hard-liners from Afghanistan, Pakistan and India get more attention in the public media, especially in the west. However, Islamic scholars maintain that Islam provides equal opportunity and respects to Muslim women. Still the reality of lives for many of them may be different from the ideal conditions depicted by scholars.

'The socio-legal status of Muslim women in contemporary Indian Society is so degraded that one wonders whether this is what the Islam scriptures have contributed to Muslim women's stability, progress and prosperity.'

'Deaf men are also better placed to find their own solutions. Deaf Afghan women can easily become very isolated at home, being unable, through social restrictions on women's mobility, to seek the company of other deaf women and being largely excluded from the social world of hearing women.'

However in spite of the differences in philosophies of different religions and the roles they may foresee for women in the society, the experience of being a women in South Asia seems to have some common sharing of cultural values, which may cut across the boundaries of different religions and geographical boundaries:

'But one common cultural image of a woman, which is neither Hindu nor Muslim, nor Christian, is that an Indian woman, though carrying the image of mother and goddess, should be in truth an embodiment of suffering, self-sacrifice and self-effacement.'

Even if South Asians migrate to other parts of the world, they may still carry the same cultural and social values. Mukhopadadhaya (p. 93, 1996) recounts her experience about a meeting on gender issues held in England:

'The Indians, Pakistanis, and Bangladeshis had united (leaving aside, for the time being, their bitter differences on the sub-continent) in a vigorous defence of culture and tradition: a tradition which respected its women, a tradition which was protective of its women, and one in which women were the centre of families which, in turn, were collectivities of co-operation, love and sacrifice.'

Given the shared perception about role of woman in the society where she is seen as "property" of a father or a husband or a son, marriage assumes a vital importance in their lives. In any discussion about disabled South Asian women, marriage, or rather difficulties in marriage are seen as the crucial issue:

Marriage is the biggest problem of women with disabilities in the country. 80 percent disabled women are unmarried.

In rural communities and also among the urban lower middle class and poorer families the wife is expected to do a lot of physical labour. So when a marriage alliance is 'negotiated' between the two families, the presence, the nature and the level of physical disability of the prospective bride assume importance for the families involved.

Some of the women were especially bitter, claiming that a double standard existed within their societies in favour of disabled men, which approved and even encouraged marriage for disabled men but found it totally unacceptable for disabled women.

INFLUENCE OF CASTE AND ETHNIC ORIGINS

The persons of tribal origins, speaking a different language and perceived as being culturally different, face discrimination in many parts of South Asia. The social divisions of caste are known mainly among Indian Hindu community. The persons defined as untouchables or as Gandhi called them, *Harijans* (God's people) or, as they prefer to call them selves, the *Dalit* (Crushed persons), face the most virulent expression of the Hindu caste oppression. However, a deeper analysis shows that similar mechanisms of systematic oppression of persons on the basis of their social origin are common in the whole region and among persons belonging to other religious groups.

'The focus on Christian *Dalits* had to contend with several objections. Two of these may be mentioned here. One was the allegation that we were dividing the Christian community...second, voiced by the church leadership and non-Christian leadership alike, and was that in Christianity there was no caste. The later is of course, sociologically and theologically absurd as an argument against Christian *dalit* struggle.'

If the *Dalit* community has to constantly fight against social discrimination, the situation of *Dalit* women is even worse. 'Caste encapsulates politics, religion, and culture in one over-arching structure that poses a formidable obstacle to constructive change. As a *harijan* leader in Tamil Nadu puts it, 'All men are born unequal and women hardly count'.

Mainstream education system, not recognising and uncaring towards the social and cultural specifics of different groups can promote, alienation of children from tribal and *dalit* groups, labelling many of them as "slow learners":

'The linguistically untrained teachers put every speech form in the pigeon hole of 'right' or 'wrong' based on what they were told is 'correct'. The *Dalit* children find that their own linguistic code is always placed in the slot labelled 'wrong'. The *Dalit* child is punished or ridiculed in the classroom for using the code acquired at home. What is ignored by the teachers is that language is an identifying label and the teacher who rejects the linguistic code of the *Dalit* child rejects not only the child's code but also all those who speak the same code, including the child's peer group, friends, and parents.' (Sreedhar, 1999)

RURAL-URBAN DIVIDES

Oppression may assume different forms and intensity in rural and urban areas. For persons living in rural communities, living conditions could mean more poverty, less services and more traditional hierarchies as compared to urban areas. At the same time, rural communities may be perceived to be more supportive compared to urban areas:

Rural communities are generally more tolerant of individual differences and unless very severely disabled, the woman is able to merge in to the family routine and household chores. But social perceptions and superstitions can greatly hinder the differently abled woman's attempts to integrate in to society. (Coleridge, p. 54, 1993)

On the other hand the discrimination against *Dalits* and tribal groups is much stronger in rural communities and the information about the oppression of disabled women from *Dalit* and tribal groups is very limited:

'There are many other problems facing women in India, especially poorest women in rural areas with whom Vigi has worked for 30 years. From birth there is discrimination against girl children expressed subtly though socialisation and more overtly though differential feeding and health care.' (O'Connell, p. 16, 1993)

'Finally, little is actually known about disabled women in rural areas. The present survey did not shed light on this aspect of the overall picture. Yet the great majority of this group lives in rural settings. How differently do they fare when compared with their city-dwelling sisters?' (ILO, p. 25, 1989)

Poverty itself, going beyond the rural-urban divide may a more significant factor in determining the powerlessness of oppressed groups:

'Second, deprivation cannot be judged only by the lowness of incomes. We have to consider the "unfreedom" involved in illiteracy, preventable morbidity, premature mortality, social exclusion and other such deprivations. These unfreedoms deserve much greater attention than they get. Regarding the urban-rural divide, this manifests itself both in terms of differences in incomes per head (typically much lower in the rural areas) and in the greater intensity of other types of deprivation, related to schooling, healthcare and so on.' (Sen, 1999)

To be a Disabled Woman in South Asia

To be born a woman in South Asia could mean a fight for survival starting from the first day of life, in which nothing can be taken for granted. For disabled female children, born in poor rural *dalit* and tribal families, the possibilities for survival become even less. If they are able to survive, their lives are more likely to be in isolation and severe oppression. Hardly any information exists in the literature about their lives and it may not be very easy to collect such information.

'Women and girls with disabilities in every community, whether urban or rural, whether in a developed or a developing country or area within ESCAP region, experience triple discrimination: from being female, disabled and poor. Discrimination and prejudice prevail even within each of these groups. Among women, the disabled women are seen as inferior; among people with disabilities, she is not their equal. Thus women with disabilities are among the most isolated, marginalised and poorest of people. They are likely to receive less care and food and be left out of family interactions, and community activities. They also have less access to health care and rehabilitation services, fewer opportunities to be educated and employed and little hope of marriage. They will be most vulnerable to physical and psychological abuse.' (UN-ESCAP, 1993)

'Disabled women face discrimination from birth. It is not unknown for children born with congenital disabilities to be killed or left to die. This is far more likely to happen in the case of a girl-child. If allowed to survive, the girl-child is likely to continue to face discrimination with the family, receive less care and food, and be left out of family interactions and activities. She will have less access to health care and rehabilitation services, fewer education and employment opportunities, and little hope of marriage. She will also be more vulnerable to mental and physical abuse.'(ESCAP - UN, p. 1, 1995)

The rural poor and the urban slum dwellers, most of them *Dalits*, often end up with precarious and life-threatening jobs. They are also forced to live in congested and unhygienic conditions. In such situations even children are expected to contribute to the family income or at least to look after their younger siblings while their parents work. Under such conditions a disabled child may be seen as an unaffordable burden which threatens the survival of whole family.

'Disability creates and exacerbates poverty, because of economic strain and isolation, not just among individual men or women, but for the disabled person's entire family. However, because far fewer opportunities for productive work or gainful employment exist for the disabled woman than for disabled man, she is perceived as posing a greater burden for the family.' (ESCAP - UN, p. 1, 1995)

Relevance of Social Model of Disability in South Asia

It has been proposed that there are cultural differences so that South Asian communities may be different from the western communities, in terms of perceptions about individual rights, role of families, etc. Thus it has been argued that theories elaborated in the developed world may not be relevant to them. Miles M. (1996, p. 488) argues that issues of individual rights promoted by western or westernised disability evangelists don't keep account of such cultural differences or of traditional mutual supporting mechanisms:

'In many such nations, traditional gemein-schaft caring continued in rural localities and kinship groups- people had few or no 'rights' as individuals, of the sort which they could 'claim' against state or society; but everyone was born into a network of relation-ships entailing mutual duties and obligations with a strong religious and economic underpinnings.'

'.. are probably reinforced by western sociological imperialism wherein the efforts of health and rehabilitation workers past and present are rubbished by air-conditioned journalists and ivory-tower social scientists, seldom having personal experience of disability service planning and delivery. The inutility to developing countries of much western social science has been documented angrily by people who have tried it.'

Similar criticism has also been made against the relevance of civil rights movements for the fight of *Dalits* or of women against their oppression. It is true that the definition of the 'rights' could be different in different cultural contexts and that different communities may decide to use their own specific ways to fight for those rights. Even concepts of equity, justice, etc. may mean different things in different cultures. Such perceptions and the strategies for fighting for equal opportunities, equity and justice could be very different from those used by oppressed groups in the developed world.

'Feminism in the West came as an offshoot of individualism _ the doctrine which holds that the interests of the individual should take precedence over the interests of the social group, family, or the state. However, in India, despite the cultural diversity among its various social, caste, and religious groups, there is a pervasive belief shared equally by men and women that individual rights must be strengthened not by pitching yourself against or isolating yourself from family and community, but rather by having your rights recognised within it. For individual rights to be meaningful, they have to be respected by those with whom you are close, rather than being asserted in a way that estranges you from them. The vast majority of Indian men and women grow up to believe that the interests of the family are primary and take precedence over individual interests.' (Kishwar, 1998)

However, to suggest that oppression and human rights are only western concepts, would negate the efforts made by women, *dalit*, tribal, disabled persons, gays and lesbians and other oppressed groups working at grass-roots level in South Asia. It may be interesting to see who are the persons who make such criticism against the 'modern' or 'western' ideas – critics are persons who, benefit from the oppressing systems?

Women and *Dalit* activists, writers, artists, social workers in South Asia have started to write about the efforts coming from the oppressed groups themselves. Pandey, feminist writer in Hindi, reacting to a statement made by a Hindu religious leader, wrote:

'It seems that while passing from the roads of Calcutta in his royal car, Shankaracharyaji kept his divine eyes closed, so that he didn't see thousands of homeless, rejected women begging on the streets, covered in just one cloth, with a half-dead child in their arms. Not that he has a good track record of worrying about the welfare of women. I don't remember if this strong supporter of building the Rama temple in Ayodhaya, has ever said anything about the oppression faced by bonded labour women, prostitutes or women forced to beg on streets.' (Pandey M., 1998, translated from Hindi).

Sen writes about use of folk songs by village women in India, where they use the stories from Hindu religious texts like Ramayana, for expressing their view about their oppression. Surely such songs are not inspired by ideas of western feminists.

'But there are always alternative ways of using a myth. If patriarchy has used the Sita myth to silence women, the village women have picked up the Sita myth to give themselves a voice.

They have found a suitable mask in the myth of Sita, a persona through which they can express themselves, speak of their day-to-day problems, and critique patriarchy in their own fashion.' (Sen, 1998)

If the reality of oppression is as true for South Asian disabled men and women, why has this not been translated in to a social movement for equal opportunities and human rights? Or rather, why has this movement remained restricted to a few urban and educated disabled persons, mostly men? Ramaiah (1998) looks at the difficulties in organising *Dalit* groups, who form 16% of the population in India and majority of them live in rural area, fragmented in small groups and isolated from each other.

'The *Dalits* are often in negligible numbers in villages with the CHs being a majority....Despite these exploitations in villages, the *Dalits* have for long, not been able to challenge the village system effectively. It is mainly because most of them are landless labourers ...'

In many ways the difficulties of rural *Dalit* groups to organise together for their rights are similar to those faced by disabled persons. If social model of disability is not seen as sufficiently relevant to South Asia, perhaps similar difficulties of isolation, fragmentation, extreme vulnerability are responsible for this? Very little information is available about the impact of oppression, the coping mechanisms or about the expression of any protest by the oppressed groups in such situations. A report by ESCAP-UN (1993), looking at the attention paid by disabled person's organisations on the needs of disabled women in Asia and pacific countries suggests:

Self-help organisations of people with disabilities should: take special action to recruit more women and girls with disabilities as members so as to improve the gender balance in the organisations and better represent the interests of women and girls with disabilities.

Only when disabled women from oppressed groups like *Dalits* become part of disabled persons' organisations and start sharing their experiences, can we learn more about their lives and their fight for survival.

Conclusions

The section looks at different factors affecting the vulnerability of disabled women in South Asia region. The paper argues that in spite of geographical, religious and ethnic differences, there are common socio-cultural patterns, based on patriarchy, which affect the role of women in the whole region.

Belonging to specific population groups like *Dalits*, which have been objects of ferocious oppression through socially and culturally sanctioned mechanisms, evolved over many centuries, creates the context for even stronger discrimination. To be a woman, and to be a disabled woman in such a situation, becomes such an extreme, where the survival itself becomes difficult. Perhaps for such a situation, 'oppression' may not be adequately strong word to describe their condition?

For any collective action, it is important not to create hierarchies or divisions of oppression. At the same time each experience of disability is unique and it is not easy to create "common cause" unless the commonality of experiences is acknowledged g in rural areas and belonging to poor families worsens the vulnerability of women.

5

Marriage in Various Religions

Marriage is a social, religious, spiritual or legal union of individuals that creates kinship. This union may also be called matrimony, while the ceremony that marks its beginning is usually called a wedding and the married status created is sometimes called wedlock.

Marriage is an institution in which interpersonal relationships (usually intimate and sexual) are acknowledged by the state, by religious authority, or both. It is often viewed as a contract. Civil marriage is the legal concept of marriage as a governmental institution, in accordance with marriage laws of the jurisdiction. If recognized by the state, by the religion(s) to which the parties belong or by society in general, the act of marriage changes the personal and social status of the individuals who enter into it.

People marry for many reasons, but usually one or more of the following: legal, social, and economic stability; the formation of a family unit; procreation and the education and nurturing of children; legitimizing sexual relations; public declaration of love; or to obtain citizenship.

Marriage may take many forms: for example, a union between one man and one woman as husband and wife is a monogamous heterosexual marriage; polygamy - in which a person takes more than one spouse - is common in some societies. Some jurisdictions and religious denominations recognize same-sex marriage, uniting people of the same sex.

A marriage is often formalized during a marriage ceremony, which may be performed either by a religious officiant, by a secular state authorized officiator, or (in weddings that have no church or state affiliation) by a trusted friend of the wedding participants. The act of marriage usually creates normative or legal obligations between the individuals involved and, in many societies, their extended families.

DEFINITIONS

Anthropologists have documented a diverse variety of marriage practices across different cultures. Many competing definitions of marriage have been proposed to capture its essential, cross-cultural characteristics.

In his three volumes of *The History of Human Marriage* (1921), Edward Westermarck defined marriage as "a more or less durable connection between male and female, lasting beyond the mere act of propagation till after the birth of the offspring."

The anthropological handbook *Notes and Queries* (1951) defined marriage as "a union of a man and a woman such that children of the woman are recognized as legitimate by both parents." Because the Nuer of Sudan allow for female-female marriage, Kathleen Gough suggested "a woman and one or more other persons." A legitimacy-based definition has been criticized as not being universal and as being circular.

Edmund Leach argued that no one definition of marriage applied to all cultures. He offered a list of ten rights associated with marriage, including sexual monopoly and rights with respect to children, with specific rights differing across cultures.

Duran Bell proposed that marriage has traditionally been characterized by sexual access rights, suggesting that modern societies which do not provide for such rights have moved toward a distinct institution.

HISTORY

Origin of marriage may lie in a man's need for assurance as to paternity of his children. He might therefore be willing to pay a bride price or provide for a woman in exchange for exclusive sexual access. Legitimacy is the consequence of this transaction rather than its motivation. In Comanche society, married women work harder, lose sexual freedom, and do not seem to obtain any benefit from marriage. But nubile women are a source of jealousy and strife in the tribe, so they are given little choice other than to get married. "In almost all societies, access to women is institutionalized in some way so as to moderate the intensity of this competition."

The way in which a marriage is conducted has changed over time, as has the institution itself. Although the institution of marriage pre-dates reliable recorded history, many cultures have legends or religious beliefs concerning the origins of marriage.

EUROPEAN MARRIAGES

For most of European history marriage was more or less a business agreement between two families who arranged the marriages of their children. Romantic loves, and even simple affection, were not considered essential. Historically, the perceived necessity of marriage has been stressed.

In Ancient Greece, no specific civil ceremony was required for the creation of a marriage - only mutual agreement and the fact that the couple must regard each other as husband and wife accordingly. Men usually married when they were in their 20s or 30s and expected their wives to be in their early teens. It has been suggested that these ages made sense for the Greek because men were generally done with military service by age 30 and marrying a young girl ensured her virginity. Married Greek women had few rights in ancient Greek society and were expected to take care of the house and children. Time was an important factor in Greek marriage. For example there were superstitions that being married during a full moon was good luck and, according to Robert Flacelière, Greeks married in the winter.

Like the Greeks, Roman marriage and divorce required no specific government or religious approval. Both marriage and divorce could happen by simple mutual agreement. There were several types of marriages in Roman society. The traditional ("conventional") form called *conventio in manum* required a ceremony with witnesses and was also dissolved with a ceremony. In this type of marriage, a woman lost her family rights of inheritance of her old family and gained them with her new one. She now was subject to the authority of her husband. There was the free marriage known as *sine manu*. In this arrangement, the wife remained a member of her original family; she stayed under the authority of her father, kept her family rights of inheritance with her old family and did not gain any with the new family. A law in the Theodosian Code (*C. Th.* 9.7.3) issued in 342 CE prohibited same-sex marriage, but the exact intent of the law and its relation to social practice is unclear, as only a few examples of same-sex marriage in ancient Rome exist.

A WOODCUT OF A MEDIEVAL WEDLOCK CEREMONY FROM GERMANY

From the early Christian era (30 to 325 CE), marriage was thought of as primarily a private matter, with no religious or other ceremony being required. Until 1545, Christian marriages in Europe were by mutual consent, declaration of intention to marry and upon the subsequent physical union of the parties. The couple would promise verbally to each other that they would be married to each other; the presence of a priest or witnesses was not required. This promise was known as the "verbum." If freely given and made in the present tense (e.g., "I marry you"), it was unquestionably binding; if made in the future tense ("I will marry you"), it would constitute a betrothal. One of the functions of churches from the middle Ages was to register marriages, which was not obligatory. There was no state involvement in marriage and personal status, with these issues being adjudicated in ecclesiastical courts.

In the 12th century, aristocrats believed love was incompatible with marriage and sought romance in adultery. Troubadors invented courtly love which involved secret but chaste trysts between a lover and a beloved. The average age of marriage in the late 1200s into the 1500s was around 25 years of age. Beginning in the 1500s it was unlawful for a woman younger than 20 years of age to marry.

As part of the Counter-Reformation, in 1545 the Council of Trent decreed that a Roman Catholic marriage would be recognized only if the marriage ceremony was officiated by a priest with two witnesses. The Council also authorized a Catechism, issued in 1566, which defined marriage as, "The conjugal union of man and woman, contracted between two qualified persons, who oblige them to live together throughout life." This change did not extend to the regions affected by the Protestant Reformation, where marriage by consent continued to be the norm. As part of the Reformation, the role of recording marriages and setting the rules for marriage passed to the state. By the 1600s many of the Protestant European countries had a state involvement in marriage.

STATE RECOGNITION

In the early modern period, John Calvin and his Protestant colleagues reformulated Christian marriage by enacting the Marriage Ordinance of Geneva, which imposed "The dual requirements of state registration and church consecration to constitute marriage" for recognition.

In England and Wales, Lord Hardwicke's Marriage Act 1753 required a formal ceremony of marriage, thereby curtailing the practice of Fleet Marriage. These were clandestine or irregular marriages performed at Fleet Prison, and at hundreds of other places. From the 1690s until the Marriage Act of 1753 as many as 300,000 clandestine marriages were performed at Fleet Prison alone. The Act required a marriage ceremony to be officiated by an Anglican priest in the Anglican Church with two witnesses and registration. The Act did not apply to Jewish marriages or those of Quakers, whose marriages continued to be governed by their own customs.

In England and Wales, since 1837, civil marriages have been recognised as a legal alternative to church marriages under the Marriage Act 1836. In Germany, civil marriages were recognised in 1875. This law permitted a declaration of the marriage before an official clerk of the civil administration, when both spouses affirm their will to marry, to constitute a legally recognised valid and effective marriage, and allowed an optional private clerical marriage ceremony.

CHINESE MARRIAGE

The mythological origin of Chinese marriage is a story about Nüwa and Fu Xi who invented proper marriage procedures after becoming married.

In ancient Chinese society, people of the same surname were not supposed to marry and doing so was seen as incest. However, because marriage to one's maternal relatives was not thought of as incest, families sometimes intermarried from one generation to another. Over time, Chinese people became more geographically mobile. Couples were married in what is called an extra-clan marriage, better known as antithetic marriage. This occurred around 5000 BC. According to modern Chinese scholars of a Marxist persuasion, society supported a matrilineal family model, therefore husbands needed to move to, and live with, their wives' families. Yet individuals remained members of their biological families. When a couple died, the husband and the wife were buried separately in the respective clans' graveyard. In a maternal marriage, a male would become a son-in-law who lived in the wife's home. This happened in the transformation of antithetic marriage into monogamy, which signifies the decline of matriarchy and the growing dominance of patriarchy in the ancient China.

SELECTION OF A PARTNER

The selection of a marriage partner may involve either the couple going through a selection process of courtship or the marriage may be arranged by the couple's parents or an outside party, a matchmaker.

A pragmatic (or 'arranged') marriage is made easier by formal procedures of family or group politics. A responsible authority sets up or encourages the marriage; they may, indeed, engage a professional matchmaker to find a suitable spouse for an unmarried person. The authority figure could be parents, family, a religious official, or a group consensus.

In some cases, the authority figure may choose a match for purposes other than marital harmony. Some of the most popular uses of arranged marriage are for dowry or immigration. In rural Indian villages, child marriage is also practiced, with parents at times arranging the wedding, sometimes even before the child is born. This practice is now illegal under the Child Marriage Restraint Act.

In some societies ranging from Central Asia to the Caucasus to Africa, the custom of bride kidnapping still exists, in which a woman is captured by a man and his friends. Sometimes this covers an elopement, but sometimes it depends on sexual violence. In previous times, *raptio* was a larger-scale version of this, with groups of women captured by groups of men, sometimes in war; the most famous example is The Rape of the Sabine Women, which provided the first citizens of Rome with their wives.

Other marriage partners are more or less imposed on an individual. For example, widow inheritance provides a widow with another man from her late husband's brothers.

Marriage Ceremony

A marriage is usually formalised at a wedding or marriage ceremony. The ceremony may be officiated either by a religious official, by a government official or by a state approved celebrant. In many European and some Latin American countries, any religious ceremony must be held separately from the required civil ceremony. Some countries - such as Belgium, Bulgaria, the Netherlands, Romania and Turkey - require that a civil ceremony take place before any religious one. In some countries - notably the United States Canada, the United Kingdom, the Republic of Ireland, Norway and Spain - both ceremonies can be held together; the officiant at the religious and civil ceremony also serving as agent of the state to perform the civil ceremony. To avoid any implication that the state is "recognizing" a religious marriage (which is prohibited in some countries) – the "civil" ceremony is said to be taking place at the same time as the religious ceremony. Often this involves simply signing a register during the religious ceremony. If the civil element of the religious ceremony is omitted, the marriage is not recognised by government under the law.

While some countries, such as Australia, permit marriages to be held in private and at any location, others, including England and Wales, require that the civil ceremony be conducted in a place open to the public and specially sanctioned by law. In En gland, the place of marriage need no longer be a church or registry office, but could also be a hotel, historic building or other venue that has obtained the necessary licence. An exception can be made in the case of marriage by special emergency license, which is normally granted only when one of the parties is terminally ill. Rules about where and when persons can marry vary from place to place. Some regulations require that one of the parties reside in the locality of the registry office.

Within the parameters set by the law of the jurisdiction in which a marriage or wedding takes place, each religious authority has rules for the manner in which weddings are to be conducted by their officials and members.

Christianity

The marriage service is a rite of passage because it marks the passage from being single to being united with another in love. The service, despite the denomination, follows a similar pattern as that shown below. The main parts of the service are as follows:

Part 4-5 The legal declaration.

The marriage service is a mixture of religious and legal ceremony, the legal part ensures that the couple to be married can actually be marries legally. It would be illegal if they were brother or sister or if they are already married. An opportunity is made for anyone to make an objection.

Part 8 The ministry of the Word

This is where the Bible is read, this marks the service out as an act of worship rather than just a legal ceremony. There is a choice of passage.

Part 10-15 The Vows

Perhaps the most important part of the service. The couple makes promises or vows in the presence of God, these are not made lightly.

Part 18-19 The Giving of Rings

The love between the man and woman is shown by the giving and receiving of rings, it shows a mutual commitment to each other.

Islam

Muslims often practice arranged marriages, this is when the parents of the girl negotiate with the parents of the man to arrange a match between them. It is the Muslim view that their parents will always seek to find good, compatible partners for their children and they may not approve of an unwise romance. The children must, however, agree to the match.

The wedding ceremony is a simple affair and the bride does not even have to go as long as witnesses are present. Red patterns are painted in Henna on her hands and feet, these have no religious significance, and they are only for decoration. After the wedding there is a party called a Walima for friends and family. This is a religious ceremony and the men celebrate separately from the women.

A Muslim wife has many rights in the marriage, these are detailed in the wedding contract she agrees to. A sum of money called a dowry is paid to the woman, she is allowed to keep this, even if the marriage fails and ends up in divorce.

Judaism

Most marriages in Judaism take place between believing Jews, this is because the child of a non-Jewish mother is not Jewish even if the father is. Some form of partner selection takes place in Orthodox Jewish families, this isn't like an arranged marriage but the two families get to know each other before the engagement is announced. During engagement sexual relationships are forbidden.

The marriage takes place under a Chuppah, a canopy held up by four poles. The Chuppah is a symbol of their new life together in their own dwelling. Both bride and groom fast (do not eat) until after the ceremony, even though a wedding is a joyous occasion it is also solemn and the fasting serves to remind them of this. *Initial Blessing-* The Rabbi recites two blessings over a cup of wine. The bride and groom then drink from the cup. *The Marriage Contract-* The Ketubah is simply a statement of the husband's intention to feed, clothe and care for his wife. The Ketubah is read to the groom before the ceremony begins.

The Ring- The groom places the ring on the bride's finger saying "with this ring you are sanctified to me according to the Law of Moses and Israel".

The Final Blessing- The Rabbi blesses the couple and then the groom breaks a glass with his foot. This calls to mind the destruction of the Temple back in the history of Judaism. Even though a wedding day is a time of joy and celebration there is still grief because of the pain of some events in Jewish history

HINDUISM

A lot of Hindu families practice arranged marriages but unlike Islam a factor called caste is often taken into account. Caste is a social group, particularly in India. Hindus are encouraged to marry into the same caste as they belong.

Hindu weddings often lavish affairs, the bride has had her skin rubbed with turmeric powder, and this gives it a glowing quality. Her sari (garment wrapped around the body) and her jewels are the best her family can afford. Red patterns are painted on her hands and feet with henna, they are symbols of her entry into a new life.

Priest will cast a horoscope for the bride and groom, this has helped them decide on the date of the wedding. The wedding itself takes place at night, when the pole star comes out, this is because the promises they make will not change, just as the pole star does not change.

Promises in a Hindu marriage are made before God, who is believed to be present in the form of a sacred fire. The wedding ceremony is led by the priest who leads the couple to take seven steps around the fire. These are taken as they walk around the fire pot four times. These steps help the bride and groom to see that in their marriage they will walk together and carry out seven duties:

1. Take the first step for strength
2. Take the second step for power
3. Take the third step for prosperity
4. Take the fourth step for happiness
5. Take the fifth step for children
6. Take the sixth step for enjoyment of pleasure
7. Take the seventh step for close union.

COHABITATION

Marriage is an institution which can join together people's lives in a variety of emotional and economic ways. In many Western cultures, marriage usually leads to the formation of a new household comprising the married couple, with the married couple living together in the same home, often sharing the same bed, but in some other cultures this is not the tradition. Among the Minangkabau of West Sumatra, residency after marriage is matrilocal, with the husband moving into the household of his wife's mother. Residency after marriage can also be patrilocal or avunculocal. Also, in southwestern China, walking marriages, in which the husband and wife do not live together, have been a traditional part of the Mosuo culture. Walking marriages have also been increasingly common in modern Beijing. Guo Jianmei, director of the center for women's studies at Beijing University, told a Newsday correspondent, "Walking marriages reflect sweeping changes in Chinese society." A similar arrangement in Saudi

Arabia, called misyar marriage, also involves the husband and wife living separately but meeting regularly.

Conversely, marriage is not a prerequisite for cohabitation. In some cases couples living together do not wish to be recognised as married, such as when pension or alimony rights are adversely affected, or because of taxation consideration, or because of immigration issues, and for many other reasons. In modern western societies some couples cohabitate before marriage to test whether such an arrangement might work in the long term.

In some cases cohabitation may constitute a common-law marriage, and in some countries the laws recognise cohabitation in preference to the formality of marriage for taxation and social security benefits. This is the case, for example, in Australia.

SEX AND PROCREATION

Marriage typically requires consummation by sexual intercourse, and non-consummation (that is, failure or refusal to engage in sex) may be grounds for an annulment. There are some married couples who remain childless either by choice or due to infertility or other factors preventing conception or bearing of children. In some cultures, marriage imposes an obligation on women to bear children. In northern Ghana, for example, payment of bride wealth signifies a woman's requirement to bear children, and women using birth control face substantial threats of physical abuse and reprisals.

On the other hand, marriage is not a prerequisite for having children, and having children outside of marriage is today not as uncommon as it used to be. In the United States, the National Center for Health Statistics reported that in 1992, 30.1 percent of births were to unmarried women. In 2006, that number had risen to 38.5 percent. Until recently, children born outside of marriage were termed illegitimate and suffered legal disadvantages and social stigma. In recent years the legal relevance of illegitimacy has declined and social acceptance increased, especially in western countries.

Many of the world's major religions look with disfavor on sexual relations outside of marriage. Sexual relations by a married person with someone other than his/her spouse is normally called adultery, and is also frequently disapproved by the major world religions (some calling it a sin), and has often been - in some jurisdictions continues to be - a crime and grounds for divorce.

MARRIAGE LAW

Rights and obligations

Marriage, by definition, bestows rights and obligations on the married parties, and sometimes on relatives as well, being the sole mechanism for the creation of affinal ties (in-laws). These may include:

- Giving a husband/wife or his/her family control over a spouse's sexual services, labor, and property.
- Giving a husband/wife responsibility for a spouse's debts.
- Giving a husband/wife visitation rights when his/her spouse is incarcerated or hospitalized.
- Giving a husband/wife control over his/her spouse's affairs when the spouse is incapacitated.
- Establishing the second legal guardian of a parent's child.
- Establishing a joint fund of property for the benefit of children.
- Establishing a relationship between the families of the spouses.

These rights and obligations vary considerably between societies, and between groups within society.

Right to Marriage

Article 16 of the Universal Declaration of Human Rights declares that "Men and women of full age, without any limitation due to race, nationality or religion, have the right to marry and to found a family. They are entitled to equal rights as to marriage, during marriage and at its dissolution. Marriage shall be entered into only with the free and full consent of the intending spouses." The Cairo Declaration on Human Rights in Islam gives men and women the "right to marriage" regardless of their race, colour or nationality, but not religion.

Marriage Restrictions

Marriage is an institution that is historically filled with restrictions. From age, to gender, to social status, restrictions are placed on marriage by society for reasons of benefiting the children, passing on healthy genes, to keep property concentrated, or (in some cases) because of prejudice and fear.

Some legal, social, or religious restrictions apply in some countries on the genders of the couple. In response to changing social and political attitudes, some jurisdictions and religious denominations now recognize marriages between people of the same sex. In some jurisdictions these are sometimes called civil unions or domestic partnerships, while some others explicitly prohibit same-sex marriages.

Societies have often placed restrictions on marriage to relatives, though the degree of prohibited relationship varies widely. In most societies, marriage between brothers and sisters has been forbidden. All mainstream religions prohibit some marriages on the basis of the consanguinity (lineal descent) and affinity (kinship by marriage) of the prospective marriage partners, though the standards vary.

There have been countless restrictions placed on marriage by different societies throughout human history, including recently. Restrictions against polygamy and marrying within a particular group or race have been common. Many societies, even some with a cultural tradition of polygamy, recognize monogamy as the only valid form of marriage. Many societies have also adopted other restrictions on whom one can marry, such as prohibitions of marrying persons with the same surname, or persons with the same sacred animal. Societies have also at times required marriage from within a certain group. Anthropologists refer to these restrictions as endogamy. An example of such restrictions would be a requirement to marry someone from the same tribe.

State Recognition

In many jurisdictions, a civil marriage may take place as part of the religious marriage ceremony, although they are theoretically distinct. Some jurisdictions allow civil marriages in circumstances which are notably not allowed by particular religions, such as same-sex marriages or civil unions.

Marriage relationships may also be created by the operation of the law alone, as in common-law marriage, sometimes called "marriage by habit and repute." The status in the eyes of one authority may not be the same as for another, e.g., a marriage may be recognised civilly, but not by a church, and vice versa.

MARRIAGE AND RELIGION

All mainstream religions have strong views relating to marriage. Most religions perform a wedding ceremony to solemnize the beginning of a marriage. Some regard marriage as simply a contract, while others regard it as a sacred institution.

Christianity

Liturgical Christian communions - notably Anglicanism, Catholicism, and Orthodoxy - consider marriage (sometimes termed *holy matrimony*) to be an expression of divine grace, termed a *sacrament* or *mystery*. In Western ritual, the ministers of the sacrament are the husband and wife themselves, with a bishop, priest, or deacon merely witnessing the union on behalf of the church, and adding a blessing. In Eastern ritual churches, the bishop or priest functions as the actual minister of the Sacred Mystery (Eastern Orthodox deacons may not perform marriages). Western Christians commonly refer to marriage a vocation, while Eastern Christians consider it an ordination and a martyrdom, though the theological emphases indicated by the various names are not excluded by the teachings of either tradition. Marriage is commonly celebrated in the context of a

Eucharistic service (a nuptial Mass or Divine Liturgy). The sacrament of marriage is indicative of the relationship between Christ and the Church yet most Reformed Christians would deny the elevation of marriage to the status of a sacrament, nevertheless it is considered a covenant between spouses before God.

In Catholicism, a principle objective of marriage is procreation: "entering marriage with the intention of never having children is a grave wrong and more than likely grounds for an annulment." Members of the Church of Jesus Christ of Latter-day Saints believe that "marriage between a man and a woman is ordained of God and that the family is central to the Creator's plan for the eternal destiny of His children." The LDS belief is that marriage between a man and a woman can last beyond death and into eternity.

Judaism

In Judaism, marriage is viewed as a contractual bond commanded by God in which a man and a woman come together to create a relationship in which God is directly involved. Though procreation is not the sole purpose, a Jewish marriage is also expected to fulfill the commandment to have children. The main focus centers on the relationship between the husband and wife. Kabbalistically, marriage is understood to mean that the husband and wife are merging together into a single soul. This is why a man is considered "incomplete" if he is not married, as his soul is only one part of a larger whole that remains to be unified.

Islam

Islam also commends marriage, with the age of marriage being whenever the individuals feel ready, financially and emotionally. In Islam, polygamy is allowed for men, with the specific limitation that they can only have up to four wives at any one time, given the religious requirement that they are able to and willing to partition their time and wealth equally among the respective wives.

For a Muslim wedding to take place, the bride and her guardian must both agree on the marriage. Should either the guardian or the girl disagree on the marriage, it may not legally take place. In essence, while the guardian/father of the girl has no right to force her to marry, he has the right to stop a marriage from taking place, given that his reasons are valid. The purpose of this practice is to ensure that a woman finds a suitable partner whom she has chosen not out of sheer emotion.

From an Islamic (Shari'ah) law perspective, the minimum requirements and responsibilities in a Muslim marriage are that the groom provide living expenses (housing, clothing, food, maintenance) to the bride, and in return, the bride must be a partner(sexually) to the husband. All other rights and responsibilities are to be decided between the husband and wife, and may even be included as stipulations in the marriage contract before the marriage actually takes place, so long as they do not go against the minimum requirements of the marriage (for example: the woman marries the man with the condition that she will not sleep with him).

In Shia Islam marriage must take place in the presence of at least 2 reliable witnesses, with the consent of the guardian of the bride and the consent of both spouses (including the girl). Following the marriage, the couple is immediately allowed to consummate the marriage does require witnesses or official statement or presence in a definite place. To create a religious contract between them, it is sufficient that a man and a woman indicate an intention to marry each other and recite the requisite words in front of a Muslim priest The wedding party can be held days, or months later whenever the couple and their families want to announce the marriage in public..

In Sunni Islam, marriage must take place in the presence of witnesses, with consent bride and the consent of both spouses (including the girl). Following the marriage they may consummate their marriage.

Bahá'í

In the Bahá'í Faith marriage is encouraged and viewed as

a mutually strengthening bond, but is not obligatory. A Bahá'í marriage requires the couple to choose each other, and then the consent of all living parents.

Hinduism

Hinduism sees marriage as a sacred duty that entails both religious and social obligations. Old Hindu literature in Sanskrit gives many different types of marriages and their categorization ranging from "Gandharva Vivaha" (instant marriage by mutual consent of participants only, without any need for even a single third person as witness) to normal (present day) marriages, to "Rakshasa Vivaha" (marriage performed by abduction of one participant by the other participant, usually, but not always, with the help of other persons).

Sikhism

In a Sikh marriage, the couple makes rounds around the holy book called *Guru Granth Sahib* four times and the holy man speaks some words from *the Guru Granth Sahib* in the form of kirtan.

Same Sex Marriage

For the most part, religious traditions in the world reserve marriage to heterosexual unions, but there are exceptions including Unitarian Universalist, Metropolitan Community Church, and Quaker, United Church of Canada and Reform Jewish congregations.

FINANCIAL CONSIDERATIONS

The financial aspects of marriage vary between cultures and have changed over time.

In some cultures, dowries and bride prices continue to be required today. In both cases, the financial arrangements are usually made between the groom (and his family) and the bride's family; with the bride in many cases not being involved in the arrangement, and often not having a choice in whether to participate in the marriage.

In Early Modern Britain, the social status of the couple was supposed to be equal. After the marriage, all the property (called "fortune") and expected inheritances of the wife belonged to the husband.

DOWRY

A dowry was not an unconditional gift, but was usually a part of a wider marriage settlement. For example, if the groom had other children, they could not inherit the dowry, which had to go to the bride's children. In the event of her childlessness, the dowry had to be returned to her family, but sometimes not until the groom's death or remarriage. Often the bride was entitled to inherit at least as much as her dowry from her husband's estate. In some cultures, dowries continue to be required today, while some countries impose restrictions on the payment of dowry.

Bride Price and Dower

In other cultures, the groom or his family were expected to pay a bride price to the bride's family for the right to marry the daughter, or dower, which was payable to the bride. This required the groom to work for the bride's family for a set period of time.

In the Jewish tradition, the rabbis in ancient times insisted on the marriage couple entering into a marriage contact, called a *ketubah*. Besides other things, the *ketubah* provided for an amount to be paid by the husband in the event of a divorce or his estate in the event of his death. This amount was a replacement of the biblical dower or bride price, which was payable at the time of the marriage by the groom to the bride or her parents. This innovation was put in place because the biblical bride price created a major social problem: many young prospective husbands could not raise the bride price at the time when they would normally be expected to marry. So, to enable these young men to marry, the rabbis, in effect, delayed the time that the amount would be payable, when they would be more likely to have the sum. It may also be noted that both the dower and the *ketubah* amounts served the same purpose: the protection for the wife should her support cease, either by death or divorce. The only difference between the two systems was the timing of the payment. It is the predecessor to the wife's present-day entitlement to maintenance in the event of the breakup of marriage, and family maintenance in the event of the husband not providing adequately for the wife in his will. Another function performed by the *ketubah* amount was to provide a disincentive for the husband contemplating divorcing his wife: he would need to have the amount to be able to pay to the wife.

Morning gifts, which might also be arranged by the bride's father rather than the bride, are given to the bride herself; the name derives from the Germanic tribal custom of giving them the morning after the wedding night. She might have control of this morning gift during the lifetime of her husband, but is entitled to it when widowed. If the amount of her inheritance is settled by law rather than agreement, it may be called dower. Depending on legal systems and the exact arrangement, she may not be entitled to dispose of it after her death, and may lose the property if she remarries. Morning gifts were preserved for many centuries in morganatic marriage, a union where the wife's inferior social status was held to prohibit her children from inheriting a noble's titles or estates. In this case, the morning gift would support the wife and children. Another legal provision for widowhood was jointure, in which property, often land, would be held in joint tenancy, so that it would automatically go to the widow on her husband's death.

Islamic tradition has similar practices. A 'mahr', either immediate or deferred, is the woman's portion of the groom's wealth (divorce) or estate (death). These amounts are usually set based on the groom's own and family wealth and incomes, but in some parts these are set very high so as to provide a disincentive for the groom exercising the divorce, or the husband's family 'inheriting' a large portion of the estate, especially if there are no male offspring from the marriage. In some countries, including Iran, the mahr or alimony can amount to more than a man can ever hope to earn, sometimes up to US$

1000,000 (4000 official Iranian gold coins). If the husband cannot pay the mahr, either in case of a divorce or on demand, according to the current laws in Iran, he will have to pay it by installments. Failure to pay the mahr might even lead to imprisonment.

MODERN CUSTOMS

In many countries today, each marriage partner has the choice of keeping his or her property separate or combining properties. In the latter case, called community property, when the marriage ends by divorce each owns half. In many legal jurisdictions, laws related to property and inheritance provide by default for property to pass upon the death of one party in a marriage firstly to the spouse and secondly to the children. Wills and trusts can make alternative provisions for property succession.

In some legal systems, the partners in a marriage are "jointly liable" for the debts of the marriage. This has a basis in a traditional legal notion called the "Doctrine of Necessities" whereby a husband was responsible to provide necessary things for his wife. Where this is the case, one partner may be sued to collect a debt for which they did not expressly contract. Critics of this practice note that debt collection agencies can abuse this by claiming an unreasonably wide range of debts to be expenses of the marriage. The cost of defence and the burden of proof is then placed on the non-contracting party to prove that the expense is not a debt of the family. The respective maintenance obligations, both during and eventually after a marriage, are regulated in most jurisdictions; alimony is one such method.

Some have attempted to analyse the institution of marriage using economic theory; for example, anarcho-capitalist economist David Friedman has written a lengthy and controversial study of marriage as a market transaction (the market for husbands and wives).

TAXATION

In some countries, spouses are allowed to average their incomes; this is advantageous to a married couple with disparate incomes. To compensate for this somewhat many countries provide a higher tax bracket for the averaged income of a married couple. While income averaging might still benefit a married couple with a stay-at-home spouse, such averaging would cause a married couple with roughly equal personal incomes to pay more total tax than they would as two single persons. This is commonly called the marriage penalty.

Moreover, when the rates applied by the tax code are not based on averaging the incomes, but rather on the sum of individuals' incomes, higher rates will definitely apply to each individual in two-earner households in progressive tax systems. This is most often the case with high-income taxpayers and is another situation where some consider there to be a marriage penalty.

Conversely, when progressive tax is levied on the individual with no consideration for the partnership, dual-income couples fare much better than single-income couples with similar household incomes. The effect can be increased when the welfare system treats the same income as a shared income thereby denying welfare access to the non-earning spouse. Such systems apply in Australia and Canada, for example.

OTHER CONSIDERATIONS

Sometimes people marry for purely pragmatic reasons, sometimes called a marriage of convenience or sham marriage. For example, according to one publisher of information about "green card" marriages, "Every year over 450,000 United States citizens marry foreign-born individuals and petition for them to obtain a permanent residency (Green Card) in the United States." While this is likely an over-estimate, in 2003 alone 184,741 immigrants were admitted to the U.S. as spouses of U.S. citizens.

Some people want to marry a person with higher or lower status than them. Others want to marry people who have similar status. Hypergyny refers to the act of seeking out those who are of slightly higher social status. In most cases, hypergyny refers to women wanting men of higher status. Isogyny refers to the act of seeking out those who are of similar status.

TERMINATION

In most societies, the death of one of the partners terminates the marriage, and in monogamous societies this allows the other partner to remarry, though sometimes after a waiting or mourning period.

Many societies also provide for the termination of marriage through divorce. Marriages can also be annulled in some societies, where an authority declares that a marriage never happened. In either event the people concerned are free to remarry (or marry). After divorce, one spouse may have to pay alimony.

Several cultures have practiced temporary and conditional marriages. Examples include the Celtic practice of hand fasting and fixed-term marriages in the Muslim community. Pre-Islamic Arabs practiced a form of temporary marriage that carries on today in the practice of Nikah Mut'ah, a fixed-term marriage contract. Muslim controversies related to Nikah Mut'ah have resulted in the practice being confined mostly to Shi'ite communities.

POST-MARITAL RESIDENCE

Early theories explaining the determinants of post-marital residenced with the sexual division of labor. However, to date, cross-cultural tests of this hypothesis using worldwide samples have failed to find any significant relationship between these two variables. However, Korotayev's tests show that the female contribution to subsistence does correlate significantly with matrilocal residence in general; however, this correlation is masked by a general polygyny factor. Although an increase in the female contribution to subsistence tends to lead to matrilocal residence, it also tends simultaneously to lead to general non-sororal polygyny which effectively destroys matrilocality. If this polygyny factor is controlled (e. g., through a multiple regression model), division of labor turns out to be a significant predictor of post-marital residence. Thus, Murdock's hypotheses regarding the relationships between the sexual division of labor and postmarital residence were basically correct, though, as has been shown by Korotayev, the actual relationships between those two groups of variables are more complicated than he expected.

CONTEMPORARY VIEWS ON MARRIAGE

Criticisms

Many people have proposed arguments against marriage for various reasons. These include political and religious criticisms, pragmatic reference to the divorce rate, as well as celibacy for religious or philosophical reasons.

Controversial views

Some views about marriage are controversial. Advocates of same-sex rights criticize the exclusion of homosexual relationships from legal and social recognition and the rights and obligations it provides. At the same time advocates of the traditional marriage movement oppose any attempt to define marriage to include anything other than the union of one man and one woman, claiming that to do so would "deprive the term of its fundamental and defining meaning."

GENDER OF DEITIES

The earliest documented religions, and some contemporary animist religions, involve deification of characteristics of the natural world. These spirits are typically, but not always, gendered. It has been proposed, since the 19th century, that polytheism arose out of animism, as religious epic provided personalities to autochthonous animist spirits in various parts of the world, notably in the development of ancient near eastern and Indo-European literature. Polytheistic gods are also typically gendered. The earliest evidence of monotheism is the worship of Aten in Egypt, the teaching of Moses in the Hebrew Bible and Zoroastrianism in Persia. Aten, Yahweh and Ahura Mazda are all masculine deities, embodied only in metaphor, so masculine rather than reproductively male.

Various 19th century scholars of comparative religion proposed that prehistoric animism worshipped nature viewed predominantly as maternal, a feature notable also within neopaganism. However, anthropological research of the early 20th century, among many pre-literate cultures, established a consensus against this theory.

In Christianity, one Person of the Trinity, the Son, is believed to have become incarnate as a human male. Christians believe that the other two Persons in the Trinity, the Father and the Holy Spirit, have never been incarnated, hence having masculine gender rather than male sex. Islam, on the other hand, has a tradition that the name Allah, like its referent, can be allocated neither grammatical nor natural gender. Masculine pronouns for Allah in the Qur'an are interpreted as generic.

The gender or genderlessness of God is a controversial issue in monotheistic and some henotheistic religions.

CREATION MYTHS ABOUT HUMAN GENDER

The creation myths of many religions contain stories about how humans came to have gender. In many stories, man and woman are created at the same time, with equal standing. One example is the creation story in Genesis 1: "And God created the man in his image. In the image of God he created him. Male and female he created them.". Most commentators interpret the parallelism to be deliberately stressing that mankind is, in some sense, a "unity in diversity" from a divine perspective and that women as well as men are included in God's image. Some argue that Genesis' gender-inclusive conception of humanity contrasts sharply with the views of gender found in older literature from surrounding cultures. Some go so far as to suggest a relatively higher status of women in western society due to Judaeo-Christian influence, and based on this verse.

In other accounts, man is created first, followed by woman. This is the case in the creation account of Genesis 2, where the first woman (Eve) is created from the rib of the first man (Adam), as a companion and helper. This version is normally cited by Jewish authorities in support of patriarchy, and likewise by Christian interpreters. A similar story appears in the Qur'an.

In Plato's *Symposium*, Aristophanes provides a humorous, mythical account to explain gender and romantic attraction. There were originally three sexes: the all male, the all female, and the "androgynous," who was half man, half woman. As punishment for attacking the gods, each was split in half. The halves of the androgynous being became heterosexual men and women, while the halves of the all male and all female became gays and lesbians, respectively.

Leadership Roles

Most religions restrict leadership to men. The ordination of women has been a controversial issue in some religions where either the rite of ordination, or the role that an ordained person fulfills, has traditionally been restricted to men because of cultural or theological prohibitions.

Beginning in the 1800s, some Christian denominations have ordained women. Among those who do not, many believe it is forbidden by 1. Timothy 2. Some of those denominations ordain women to the diaconate, believing this is encouraged by 1 Timothy 3-4. Some Islamic communities (mainly outside the Middle East) have recently appointed women as imams, normally with ministries restricted to leading women in prayer and other charitable ministries.

THEOLOGICAL FIGURES

The key theological figures of all major religions have been men. However, nearly all these religions and many smaller ones, have included notable female figures from time to time.

In Indic religions, we do not know the gender of the Vedic writers, though the first hymn of the Rgveda is addressed to the masculine deity Agni, and the pantheon of the Vedas is dominated by masculine gods. The traditional religious leaders of Jainism and avatars of Vishnu are all men. Siddhartha Gautama (the Buddha) was a prince, so also a man.

In Abrahamic religions, Abraham himself, Moses, David and Elijah are among the most significant leaders documented according to the traditions of the Hebrew Bible. John the Baptist, Jesus and his apostles, and Saul of Tarsus again give the New Testament an impression of the founders and key figures of Christianity being male dominated. They were followed by a millennium of theologians known as the Church Fathers. Islam was founded by Muhammad, and his successor, Abu Bakr, was also a man.

On the other hand, The Virgin Mary, the mother of Jesus of Nazareth, is not associate with leadership or teaching, but is to the founding of the Seventh-day Adventist Church and is officially considered a prophet by Seventh-day Adventists. In particular, White's biblical commentaries and other writings are often considered inspired or even infallible. Mary Baker Eddy was the founder of the Christian Science movement.

Segregation

In traditional Jewish synagogues, the women's section is separated from the men's' section by a wall or curtain called a mechitza. Men are not permitted to pray in the presence of women, to prevent distraction. The *mechitza* shown in the picture on the right is one in a synagogue affiliated with the 'left wing' (more modern side) of Modern Orthodox Judaism, which requires the *mechitza* to be of the height shown in the picture. More traditional or 'right wing' Modern Orthodox Judaism, and all forms of Haredi Judaism, requires the *mechitza* to be of a type which absolutely prevents the men from seeing the women.

Enclosed religious orders are usually segregated by gender. Sex segregation in Islam includes restrictions on interaction between men and women. Men and women also worship separately in mosques.

ROLES IN MARRIAGE

Nearly all religions recognize marriage. Many have views on appropriate gender roles within marriage. Two notable views are Complementarianism and Christian Egalitarianism.

In Genesis 3, Adam names his wife Eve (Hebrew Chayyah) because she will be "mother of all the living".

In Islam a woman's primary responsibility is usually interpreted as fulfilling her role as a wife and mother, whereas a man's role is to work and be able to financially support his wife and family.

Abortion

In most religions abortion is considered to be immoral, the taking of God's gift of life to a child, for the sake of self-interest. The rationale for this view often also involves appeals to mothering responsibilities and privileges considered to be divine gifts to exclusively given to women.

Homosexuality

Homosexuality is expressly forbidden in many religions, but typically in casuistic rather than apodictic laws. As such, the rationale for such proscriptions is not clearly evident, though avoidance of procreation and contribution to society via establishing families are sometimes offered as pragmatic considerations.

Feminism and Religion

Feminism has engaged with religion in three main ways:

- From *outside* religions, often critically, feminist scholars have cited religious texts and practice as evidence of patriarchal institutions within humans societies, antithetical to feminist aims;
- From *within* religions, some members, scholars and leaders sympathetic to feminism have attempted reforms to harmonize contemporary religious thought and practice to be more in line with feminism; and,
- Independently of established religion, some feminist thinkers have attempted to construct a theology or range of religious practices based on feminist principles, notably but not restricted to Wicca, neopaganism and various New Age religions.

Several theologians have considered women as morally inferior to men, and many have considered women the source of temptations, especially sexual temptations, for men. In traditions where God is considered male, there is sometimes a view that men are more like God than women. For example, in 1 Corinthians 11:7, Paul asserts that "man is the image and glory of God: but the woman is the glory of the man". In some case leading to a view of male superiority and female inferiority.

Feminist Theology

Feminist theology is a movement generally in Christianity and Judaism to reconsider the traditions, practices, scriptures, and theologies of their religion from a feminist perspective.

Likewise, those who practice feminist spirituality may adhere to a feminist re-interpretation of Western monotheistic traditions. In these cases, the notion of God as having a male gender is rejected, and God is not referred to using male pronouns.

Religious ideas about gender have sometimes supported gender equality and a relatively higher status of women. Pierre Chaunu has argued that the influence of Christianity is the main factor leading to equality for women. Priyamvada Gopal argues an interesting perspective: that gender equality is indeed a product of Judeo-Christian doctrine, but not exclusive to it. He expresses concern that gender equality is used by western countries as a rationale for "neocolonialism". Jamaine Abidogun argues another interesting perspective: that Judeo-Christian influence has indeed shaped gender roles in Nigera (a strongly Christianised country), however, she doesn't consider feminism to be a product of Judeo-Christian doctrine, but rather a preferable form of "neocolonialism".

"Pierre Chaunu also contends that the very recognition of women as full fledge human beings depends on moral and ethical categories that derive ultimately from the Bible. He points out, for example, in his book *Foi et histoire* (Faith and History, 1980), that it is only in those cultures where the biblical text and Christianity have had some long term influence, that the status of women has gradually improved from that of property and progenitor to that of a full human being, equal to man. In cultures where the biblical text has not had any significant impact, women are regarded as property whose main purpose is to produce children. In such cultures, women are married as soon as they are able to procreate, they have little or no access to formal education, and they are allowed little self-determination. The main reason behind this social transformation is fundamentally linked to a statement found in Genesis 1:27: 'God created man in his own image, in the image of God he created him; male and female he created them.' This text affirms the intrinsic dignity of both men and women regardless of their gender or social status. This is in stark contrast to ancient Egypt and Mesopotamia where the value of human beings was primarily determined on the basis of their social standing." Pierre Gilbert, *How the Bible Shapes Our World*

"The insistence that equality is a western concept to be defended against the incursions of others relies on a continued deafness to resistant voices from outside Judaeo-Christian contexts. This, ironically, makes the self-proclaimed liberals who insist on these useful collaborators for authoritarian chauvinists from outside the west. For they are all in curious agreement that women's equality is a western concept and call for it, accordingly, to be either enforced (that's why we sent in the troops) or rejected (by keeping women secluded)." Priyamvada Gopal, "West has no monopoly on battle for gender equality", Kuwait Times September 30, 2007.

"These perceptions demonstrate a pattern of gender roles shaped by Western Judeo-Christian doctrine within the formal education curriculum, minimal inclusion of local history or cultural content, and loss of indigenous knowledge and practices. Gender-role change is one aspect of a general Westernizing effect of formal models of Western education on indigenous cultures." Jamaine Abidogun, "Western education's impact on Northern Igbo gender roles in Nsukka, Nigeria", *Africa Today* (2007).

6

Feminism
An Introduction

(RADICAL/CULTURAL FEMINISM AND MARXIST/ SOCIALIST FEMINISM)

RADICAL/CULTURAL FEMINISM

Radical feminism focuses on patriarchy as a system of power that organizes society into a complex of relationships producing what radical feminists claim is a "male supremacy" that oppresses women. Radical feminism aims to challenge and to overthrow patriarchy by opposing standard gender roles and what they see as male oppression of women, and calls for a radical reordering of society. Early radical feminism, arising within second-wave feminism in the 1960s, typically viewed patriarchy as a "transhistorical phenomenon" prior to or deeper than other sources of oppression, "not only the oldest and most universal form of domination but the primary form" and the model for all others. Later politics derived from radical feminism ranged from cultural feminism to more syncretic politics that placed issues of class, economics, etc. on a par with patriarchy as sources of oppression.

The term radical in radical feminism (from Latin, meaning root) is used as an adjective meaning of or pertaining to the root or going to the root. Radical feminists locate the root cause of women's oppression in patriarchal gender relations, as opposed to legal systems (liberal feminism) or class conflict (socialist feminism and Marxist feminism).

THE ROOTS OF RADICAL FEMINIST MOVEMENT

In the United States, radical feminism developed as a response to some of the perceived failings of both New Left organizations such as the Students for a Democratic Society (SDS) and liberal-feminist organizations such as the NOW. Initially concentrated mainly in big cities like New York, Chicago, Boston, Washington, DC, and on the West Coast radical feminist groups spread across the country rapidly from 1968 to 1972.

In the United Kingdom, feminism developed out of discussions within community based radical women's organizations and discussions by women within the Trotskyist left. Radical feminism was brought to the UK by American radical feminists and seized on by British radical women as offering an exciting new theory. As the 1970s progressed, British feminists split into two major schools of thought: socialist and radical.

As a form of practice, radical feminists introduced the use of consciousness raising groups (CR groups). These groups brought together intellectuals, workers and middle class women in developed Western countries to discuss their experiences. During these discussions, women noted a shared and repressive system regardless of their political affiliation or social class. Based on these discussions, the women drew the conclusion that ending patriarchy was the most necessary step towards a truly free society. These consciousness-raising sessions allowed early radical feminists to develop a political ideology based on common experiences women faced with male supremacy. Consciousness raising was extensively used in sub-units of the National Organization For Women (NOW) during the 1970s.

Within groups such as New York Radical Women (1967-1969), a radical feminist ideology began to emerge: "the personal is political" and "sisterhood is powerful," formulations that arose from these consciousness-raising sessions. New York Radical Women fell apart in early 1969 in what came to be known as the "politico-feminist split" with the "politicos" seeing capitalism as the source of women's oppression, while the "feminists" saw male supremacy as "a set of material, institutionalized relations, not just bad attitudes." The feminist side of the split, which soon began referring to itself as "radical feminists" soon constituted the basis of a new organizations in 1969, Red-stockings. founded by Shulamith Firestone and Ellen Willis in 1969.... At the same time, Ti-Grace Atkinson led "a radical split-off from NOW", which became known as The Feminists (1968-73). A third major stance would be articulated by the New York Radical Feminists, founded in 1969 by Shulamith Firestone (who broke from the Red-stockings) and Anne Koedt. From Wikipedia.

RADICAL FEMINIST THEORY

Radical feminism describes sexual oppression (patriarchy) as the fundamental form of oppression and the primary oppression for women; men as a group are considered to be the beneficiaries of this social system of male supremacy (patriarchy). Radical feminism offers a real challenge to and rejection of the liberal orientation towards the public world of men. Indeed it gives a positive value to womanhood rather than supporting a notion of assimilating women into arenas of activity associated with men. Radical feminism pays attention to women's oppression as women in a social order dominated by men. According to this approach, the distinguishing character of women's oppression is their oppression as women, not as members of other groups such as their social class. Hence, the explanation for women's oppression is seen as lying in sexual oppression. Women are oppressed because of their sex.

The notion of shared oppression is intimately connected with a strong emphasis on the sisterhood of women and encourages some degree of separatism from men. Furthermore, this identification with women and rejection of male dominance involves both a critique of the existing organization of heterosexuality as prioritising men and recognition of lesbianism as a challenge to that priority. Radical feminism stresses that in a social order dominated by men the process of changing sexual oppression must, as a political necessity, involve a focus on women. And because radical feminism recommends putting women first, making them the primary concern, this approach is inclined to accord lesbianism an honoured place as a form of mutual recognition between women.

Radical feminism's strong interest in recovering or discovering positive elements in femininity (asserting in essence that it is good to be a woman and to form bonds with other women), in combination with its location of men as the beneficiaries of sexual power relations, results in a relatively sharp division drawn between men and women. Radical feminists usually present a historically continuous, clear-cut difference between men and women. Sometimes, this is argued to be the result of an ontological (essential, intrinsic, innate) difference. However, other radical feminist writers note that male domination is a social structure and not the consequence of some in-built male propensity, even if motivations towards mastery are typically male. In other words, feminists in this tradition see a difference between men and women as inevitable (given by nature) or at least as so established historically that it is very deeply embedded.

Since radical feminist thinkers consider sexual oppression to be profoundly entrenched, frequently depicting it as the original form of coercive power, they also present the social and political changes required to overthrow the system of male domination as far-reaching. As you would expect given the name, radical feminism generally advocates a revolutionary model of social change. However, the proposed revolutionary change in the organization of power relations between the sexes is not described in terms of a single cataclysmic moment, but rather as the consequence of the cumulative effect of many small-scale actions. Moreover, revolutionary practice-conceived as the basis of radical feminist theory-is undertaken with an emphasis on small group organization rather than formalised centrally administered structures.

Radical feminist may pursue a revolutionary agenda but, like liberal feminists, they stress practical political strategies. Nevertheless, in contrast to liberal feminist framework, radical feminism is inclined to be suspicious of government intervention, perceiving the state itself as being intrinsically patriarchal, and also tends to focus on the politics of the private sphere, in particular sexuality, motherhood and bodies. Their focus on the body as a critical site of oppression for women but also as representing women's difference and therefore to be celebrated, stands in sharp contrast to liberal feminism's general aim of reducing or preferably eradicating attention to bodies and bodily difference as politically retrograde.

Radical feminists are far less unidimensional regarding the body, ranging over sexuality, sexual violence, the (maternal) reproductive body, the feminine body as a source for creativity and spirituality, and the meaning of an embodied self (feminine subjectivity and identity). Indeed, radical feminism conceives the body-and, in particular, the sexually specific body-as critical to social analysis. Sexual difference (evident in, for example, women's capacity to give birth) is not socially insignificant. Rather than perceiving the (sexed) body as mere, inanimate meat separate from social practices, power relations or social change, this form of feminism stressed the interconnection between bodies and society. The agenda of radical feminist writings is to counter women's supposedly natural, biological, inferiority and subordination within patriarchal society by asserting their at least equal (or superior) status in relation to men: a crucial aspect of that agenda is for women to gain control over their own bodies/biology and relatedly to value and celebrate women's bodies. Many aspects of radical feminism's emphasis on body politics have been taken up with enthusiasm by emerging groupings of feminists, such as psychoanalytic and postmodern/poststructuralist feminists. (Beasley)

FEMINIST SEPARATION

Only the elimination of patriarchy and the destruction of male control will liberate women. Radical feminists suggest we begin by eliminating gender, specifically sexual status, role, temperament, and social constructions as they have been constructed under patriarchy. As patriarchy is organized through men's relationships with other men, unity among women is the only effective means for liberating women. Separation by women calls for isolation from men and from institutions, relationships, roles, and activities that are male-defined and male-operated, benefit males, and maintain male privilege. Female separation often falls into the trap of essentialism, the belief in natural or innate differences between men and women.

TWO CAMPS OF RADICAL FEMINISM

With the emergence of so-called essentialism in feminist thought, radical feminists have divided into to camps: radical-libertarian feminists and radical-cultural feminists. Radical-libertarian feminists generally espouse the ideas of 1960s and 1970s radical feminists who drew attention to the ways in which the concept of femininity as well as women's reproductive and sexual role and responsibilities often limit women's developments as full human persons. These radical feminists longed for androgyny. Although this concept of androgyny is flawed due to its male bias, it captured radical feminists' desire to transcend the limits of the sex/gender system by daring to be masculine as well as feminine.

Later, some radical feminists questions whether they wanted to be masculine at all. They believed a "Bitch" was not a full human person, but a woman who had embraced some of the worst features of masculinity. These radical feminists rejected androgyny as a desirable goal, replacing it with a proposal to embrace women's essential "femaleness." Instead of believing liberated women must exhibit both masculine and feminine traits and behavior, these radical-cultural feminists expressed the view that it is better to be female/feminine than it is to be male/masculine. Thus, women should not try to be like men; they should try to be more like women, emphasizing values and virtues culturally associated with women and deemphasizing values and virtues culturally associated with men.

Lesbian feminism and cultural feminism are two types of feminist separations advocating the creation of a woman-identified world through the attachments women have to each other. The emergence of lesbian separatism positions lesbianism as more than a purely personal decision, as an outward sign of an internal rejection of patriarchal sexuality. Lesbianism becomes a paradigm for female-controlled female sexuality, the kind of sexuality that meets women's own needs and fulfills their own desires.

Another popular strategy for resisting patriarchy has been to redefine social relations by creating women-centered cultures that emphasize the positive capacities of women by focusing on the creative dimensions of their experiences. Radical feminists, more than any other feminist theoretical tradition, have made considerable contributions to women's culture. Feminist art, spirituality, food ecology, reproduction, mothering, gender, and sexuality have all been nurtured in women-only spaces.

CRITIQUES OF RADICAL FEMINISM

Radical feminists fall into the trap of essentialism, the conviction that men are men and women are women and that there is no way to change either's nature. Such a trap represents an analytic dead end as well as a political danger. Agreeing, for example, that women are by nature compassionate and nurturing and that men are aggressive and competitive suggests that radical feminists accept and promulgate the very stereotypes they are trying to avoid. How does one alter or change that from which one cannot escape? Women are trapped inside their bodies and biology becomes their destiny. From Elliot and Mandell.

MARXIST FEMINISM

Marxist feminism is a sub-type of feminist theory which focuses on the dismantling of capitalism as a way to liberate women. Marxist feminists see contemporary gender inequality as determined ultimately by the capitalist mode of production. Gender oppression is class oppression and women's subordination is seen as a form of class oppression which is maintained (like racism) because it serves the interests of capital and the ruling class. Marxist feminists have extended traditional Marxist analysis by looking at domestic labour as well as wage work in order to support their position.

MARXISM AND FEMINISM

Marxism sees class division rather than gender as the root of women's oppression. Because a great deal of the time and effort needed to reproduce the workforce comes from the private family, Marxism sees this as the key to women's oppression: despite education and healthcare in advanced capitalist countries, the system largely depends on women's unpaid labour for the creation of a healthy workforce. Women's generally lower economic and social status grows out of this. In the classic Marxist statement of this position, Friedrich Engels's The Origin of the Family, Private Property and the state (1845) argues that the bourgeois family rests on a material foundation of inequality between husband and wife, where the latter is a kind of unpaid prostitute, producing heirs for the transmission of property in exchange for board and lodging.

From this position comes The Communist Manifesto's (1848) call for the abolition of the bourgeois family, an aim that chimes in with the aims of radical feminism. For Marxists, much feminism is a bourgeois theory that seeks to reform the system to the advantage of some women, rather than get rid of the system that exploits the vast majority of women and men; see for example Naomi Wolf's call for women to enter the workforce as bosses and as workers in Fire With Fire (1993). For Marxists this would result in one class of women bosses exploiting another group of female wage labourers. Nevertheless, feminism sees Marxist economic theory as often blind to gender issues and argues that Marxism has yet to explain with reference to the needs of capitalism how women's oppression seems to exist in all known societies. There are, of course, women who are Marxist feminists and who are critical of Engels but who broadly agree with his analysis. (Gamble 269)

THEORY OF MARXIST FEMINISM

In Marxist feminism, following the work of Karl Marx, hierarchical class relations (built on unequally distributed or owned sources of wealth, including monetary and other resources) are seen as the source of coercive power and oppression, of all inequalities ultimately. Sexual oppression is seen as a dimension of class power. In this model the earliest forms of class division historically give rise to male dominance; class oppression predates sex oppression. The emerging organisation of the first forms of private wealth, and therefore of class hierarchy, led to the treatment of women as property. In other words Marxist feminism offers a version of history and society which is in some ways the opposite of that proposed by radical feminism. (In radical feminism the earliest forms of male domination over women produce a framework of hierarchical social relations in which class divisions arise; sexual oppression predated class power.) Clearly what is at stake in this difference of views is the question of which is the primary oppression for women, and hence which should be given the highest priority in feminist political struggle.

By comparison with radical feminism there is typically less concern within Marxist feminism with ideas and attitudes and more of a focus on labour and economics when exploring women's positioning. Since labour is viewed as fundamental to all economic activity, (historically specific) analysis of the organisation of labour is crucial to Marxist feminist approaches. Indeed, the organisation of labour and the tools/ technologies associated with labour are perceived in concert as constituting the underlying economic structure or system of society. This economic structure conditions the form of all other social relations in that society and in this sense is the basis of society. Hence Marxist feminists, in common with other Marxists, generally accept some version of what is called the base-superstructure model of society, that is, social relations-including those related to sexual inequality-are conceived as crucially shaped by the economic base of society, rather than by ideas and attitudes.

The Marxist feminist approach tends, like liberal feminism, to be oriented towards the public sphere and, given its concern with the organisation of labour, generally pays particular attention to women's position in relation to waged labour. The significance of unpaid labour undertaken in the private realm, which is very much associated with women, is controversial in Marxist feminism because Marxism largely equates the economy with the capitalist market-place. However, unlike liberal feminists, Marxist feminist thinkers are deeply antagonistic to the capitalist economy and advocate a revolutionary approach in which the overthrow of capitalism is viewed as the necessary precondition to dismantling male privilege.

The main enemy in this form of analysis is the class system (capitalism, in modern societies) which creates divisions between men and women. Marxist feminism shares with liberal feminism (both are what is described as equality or egalitarian feminisms), an assumption that there is an underlying sameness between men and women. While women seem to be oppressed by the men around them, they-like men-are ultimately oppressed by capitalism, and hence the interests of men and women are not crucially different. (Beasley 59-61)

SOCIALIST FEMINISM

Marxist feminism was an influential school of Western feminist thought in the 1960s and 1970s. While the impact of Marxism on feminist theory remains evident in a number of contemporary approaches, the Marxist feminist tradition is waning. Its place in advocating the significance of class analysis for feminism has now largely been overtaken by a range of socialist feminisms. Socialist feminism focuses upon both the public and private spheres of a woman's life and argues that liberation can only be achieved by working to end both the economic and cultural sources of women's oppression. Socialist feminism is a dualist theory that broadens Marxist feminism's argument for the role of capitalism in the oppression of women and radical feminism's theory of the role of gender and the patriarchy. (Beasley 58).

Debates between radical feminists and Marxist feminists in the 1960s and 1970s concerning the fundamental cause of social inequality were important in the formation of new groupings of socialist feminism. Socialist feminists attempt to maintain some elements of Marxism regarding the significance of class distinctions and labour while incorporating the radical feminist view that sexual oppression is not historically a consequence of class division. In other words all socialist feminists assert, along with radical feminists, that women's subordination predated the development of class-based societies and hence that women's oppression could not be caused by class division. There are several versions of socialist feminism which involve different combinations of radical and Marxist feminism, and which sometimes incorporate the influence of psychoanalytic feminisms.

In brief, three major socialist feminist traditions may be described as deriving from debates between radical and Marxist feminists. The first strand involves a concern with the social construction of sex (gender) which was largely seen in terms of Freudian psychoanalysis. This approach tends not to perceive sexual oppression through the Marxist lens of women's unequal socio-economic position or so-called material organisation of social life-but rather conceives that oppression as the effect of psychological functions. At the same time the approach continues to make use of a Marxian understanding of class relations. In broad term a psychological model of sexual power is presented alongside an (historically specific) economically based account of class power. Moreover, the former is moulded or historically contextualised by the organising force of the latter. Because the overall model makes use of Marxist materialism (that is, a methodology which sees economics as the fundamental motor of social relations-shaping the form of society), it tends to adopt a version of the Marxist base-superstructure model in which class is still ultimately fundamental (base) since sex/gender is psychological (superstructure). Hence, in some ways this is more a two-tier, rather than a mutual or dual, theory of social relations. The two-tier approach is epitomised by the early work of Juliet Mitchell.

The second major strand of socialist feminism attempts to draw the work of radical and Marxist feminists into one theory of power and describes a unified system sometimes referred to as capitalist patriarchy (although this term is also used by other feminists, including other socialist feminists). Examples of this approach include work by Alison Jaggar and Iris Young. These theorists of unified system oppose dual systems by attempting to analyze patriarchy and capitalism together: for example, Jaggar in *Feminist Politics and Human Nature* (1983) argues that the concept of alienation is a category broad enough to encompass both phenomena.

By contrast, the third strand describes a dual systems model of social organisation. Dual systems theorists regard patriarchy and capitalism as distinct entities, each with its own agenda and sphere of social relations. Hence the first strand of socialist feminism offers what has been termed a dual systems model of social analysis, investigating sex and class power according to different procedures and identifying two systems of social organisation corresponding to these forms of power, that is, patriarchy and capitalism. However, in the third strand both sex and class power have a material aspect, that is, they both are conceived as having an economic form. In other words, patriarchy is not seen as simply psychological, as is the case in the first variant associated with Juliet Mitchell who approaches patriarchy as an ideology that transcends the material sphere. The third form of socialist feminism offers a more full-blown account of both systems in which sexual oppressions and class oppressions interact but are not cast as dependent form. Neither is viewed as more fundamental than the other in the overall shaping of social relations. The work of Heidi Hartmann provides the classic example of dynamic duo approach.

These versions of socialist feminism are identified by their views of the relationship between class and sex (sometimes referred to as the category, gender)-that is, the relationship between capitalism and patriarchy. Other categories of power such as race tended to be marginalised in initial accounts of debates among socialist feminists. Indeed the issue of race and/or ethnicity, for example, increasingly became a point of contention within socialist feminism given its concern with forms of power that cut across both class division and sexual difference. Recently, such debates have contributed to the development of certain postcolonial feminist perspectives, indicating ongoing inter-actions between socialist feminist themes and feminist concerns regarding race/ethnicity. (Beasley 62-64)

Oppression of Women in Victorian Society

To understand the Victorian Society one needs to grasp the background and views of these times. Women were only there for men, and the most important aspect of that was marriage. From the time they were young they were set to be beautiful, because only the pretty girls got the richest man. The girls were groomed and basically put on show like a common day racehorse. In addition to being good-looking, a young girl was expected to be able to sing, play an instrument and speak French and Italian. They needed to be gentle, innocent and ignorant of their intellectual opinions. Women in the Victorian society were meant to be weak, helpless, fragile and unable to think for themselves. There only job was to make sure the home was comfortable for the children and the man of the house. Her one use to the world was to produce a large family and a neutral and calm family atmosphere. In those times the man was the moneymaker and the woman was there to keep the house clean and tranquil.

Men also held no respect for their wives with the evidence being mistresses. Many men kept mistresses and in the end still expected their wives to be faithful even though they were cheating on them the whole time. It was a hypocritical aspect of the Victorian ages; men were allowed and even praised for being unfaithful while women were ostracized for it.

A woman could not hold property in her name, even if she inherited that property from her family it was still in sole possession of her husband. If a married couple had been separated for a period of time the woman had no rights to see her own children, and if a divorce was in affect the woman had no chance of acceptance in society again.

Women were also expected of certain things according to their rank in the social class. For example, a woman in the wealthy upper class was supposed to read, sew, receive guests, write letters see to the servants and dress for the part as her husbands' social representative (Fashion-Era- "A woman's place"). A day in the life of a wealthy woman involved that of etiquette rules, which involved changing up to six times a day. There was a morning dress, walking dress, visiting dress, traveling dress, shooting dress, golf dress, seaside dress, concert dress, opera dress, dinner and ball dress (Fashion Era- "A woman's place). All the different dresses were for the sole purpose of showing off the wealth of the man and the household. In the lower classes people were more concerned with where they would be getting their next meal, eating food such as rancid bacon, old meat and tainted cheese. The social aspect of the pressure on woman to be perfect and unheard was not of great importance to these people.

A variety of rules were also written during the Victorian times, these compiled rules were placed together and labeled as etiquette books. They covered a wide range of topics, and each sex, male and female, had a separate book. These rules had to be followed by the women and men within the society or they would be viewed as complete outcasts. The following rules were the ones stated for the women they read as such:

- A lady should always walk in an easy, unassuming manner, neither looking left nor right. If anything in a store window attracted her attention, she may stop and examine it and then continue her walk (Etiquette 1).
- A lady does not giggle of whisper in public. Talking loudly is "inexcusable" at all times. She does not call her friends across the street or ask of their heath in a boisterous manner (Etiquette 1)
- When bowing on the street, the lady should merely incline

her head gracefully, never her body. She should also smile pleasantly (Etiquette 1)

- She must never let a male talk to her in a slangy manner, touch her on the shoulder of call her by her first name in front of strangers (Etiquette 1).

A woman in the Victorian Society was also not to be rude, nor attract any attention to her. She was to be kind to everyone around her, not drag her skirt on the ground while she walking, not look back at anyone whom she has passed, she was not to stare at any other woman's clothing, stop to talk to a person upon walking, and was never allowed to remove her bonnet shawl or gloves if she was visiting someone for only a short amount of time.

Even though women were held to such restraints and guidelines a few of them struggled to remain independent and fought for their rights. A quote from the book *Turning the Tide* is what I'll leave this paragraph with. To be a woman...what is that? In the discovery to this question was to lie another struggle, another battleground. Wearily but without complaint, the world army of women shouldered arms and marched forward once again.

Within the novels the women have taken the situations they were placed in and reacted in two diverse ways. One character held to her beliefs and her morals, where the other woman acted in situations as she felt would benefit herself and makes her happy even if it was for just that one moment. Jane Eyre is a woman of strength, power and feminism. Emma Bovary is weak, self absorbed and sullen throughout her story. It is my focus to show how each character examined and adapted to the Victorian society.

This passage basically describes everything about this character. She understands the society she is in, but is unwilling to do anything about it. In that passage she is talking of herself, she was forced into marriage and now lives a life she wishes she never took in the first place. In the end she takes her life to exit out of the pain she has caused herself.

Emma and Jane are diverse in many ways. Jane was the strong and passionate character, while Emma was the poster girl of a Victorian woman. Each reacted in a different way to their environments. I have given you background information and specific example of women depicted during these times. Which one would you be more like? Would you have followed the pack or set forth a new path? That is the ultimate question.

MEN AND HOME

Family roles continue to change, especially in today's over-scheduled, stressful environment. The male dominating ambience at home was a natural scene for centuries. The family structure in which man was the financial supporter and woman was the caregiver has become, to a large extent, a myth. On the domestic front, while women have taken on an increasing role in providing income to their families, men have not taken up their share of responsibility in the family life. However recent studies prove that men also have initiated to play changed roles at home.

Men were the directors and managers of household activities for a long time. Nevertheless the changes in the management skills seemed to change men's attitudes in the household affairs too. Today's man incorporates the ideas and suggestions of their female counterparts in household decisions for a better change. They do not find the kitchen as woman's world alone. They don't loathe washing or cleaning as woman's business alone. Those men who possess the old ideas with them are far outdated in the society and face serious dilemmas in their domestic life.

Man enacts an essential role during the childbirth, which is the most happy and major incident that can happen in a family. Man's role in childbirth is becoming more involved than it has been in the past. More and more women want their husbands to give active assistance during pregnancy and childbirth. Many men join classes to help their partners prepare for childbirth. They also give moral support during childbirth by simply being present during their wife's labor. However men are advised not to enter the labor room if they are fearful or anxious. The fear will be transmitted to his partner and this increases her adrenaline levels leading to labor stalling.

Men's changed role as father in the family varies all across the world. Today's man has a close relationship with his children than what his father had with him. A research from U.S.Department of Education Study indicates that the role of fathers in school and childcare promotes the achievement of their children. Men do not shy away from teaching their children societal norms, gender equality and responsible sexual relations while actively involving himself in family affairs. Thus man's active assistance in childbirth and childcare poses a great challenge to traditional notions of fatherhood held by both men and women.

Man's support in domestic activities considerably reduces the violence against women. However it must be taken into consideration that men's participation in family affairs is not complete till date. The underlining fact is that most men want to do their part well with raising children to become successful adults and any extra help, guidance, support and reinforcement for their families. And, the best part is that children benefit through happier families where dads and moms are involved productively in their lives.

FAMILY LIFE DURING THE RENAISSANCE

Typical family life in the European Renaissance consisted of parents and their children. Women were expected to stay home, while men worked in fields or in merchant shops

A typical Renaissance family consisted of parents and their children. It is traditionally assumed that people of the Renaissance married early. This is because nobility did indeed marry early. However, most common folks married in their mid twenties. Since the average life expectancy was 40, few people made it to grandparent status.

FAMILY ROLES

Renaissance women, as expected, stayed home to care for the children and the home. The men in the family worked out of the house, tending the fields or in merchant shops. Older children, both boys and girls, often worked alongside their fathers, to help contribute to the family funds.

Renaissance Housing

Houses were small, some with no more than two rooms with low ceilings. It would not be considered unusual for a family of four to share a single bed. Fireplaces were used for cooking, causing the tiny house to quickly fill with smoke. In the hot summer months, houses in southern Europe would be unbearably hot, causing women to set up their looms outdoors, and serve family meals outdoors as well.

Household possessions were simple, to say the least. A typical Renaissance family owned two outfits and one pair of shoes per person, some kitchen utensils, the aforementioned bed, a chamber pot, a table and bench, a chest, and a few tools.

Childhood

Children during the Renaissance were thought to be miniature adults, and therefore, as soon as they were out of nappies they would be dressed like adults, and spoken to as adults. Despite this grown-up treatment, children still had childhood toys and games to keep them occupied until they were deemed old enough to work, at around age seven or eight.

Bibliography

American Association of University Women. *Shortchanging Girls, Shortchanging America: A Call to Action*. Washington, DC: American Association of University Women, 1991.

Anne E. Beall and Robert J. Sternberg. *The Psychology of Gender*. New York: Gilford Press, 1993.

Anne Fausto-Sterling. *Myths of Gender: Biological Theories about Men and Women*. New York: Basic Books, 1985.

Bandura, Albert. *Social Learning Theory*. Englewood Cliffs, NJ:Prentice-Hall, 1977.

Beal, Carole R. *Boys and Girls: The Development of Gender Roles*. New York: McGraw-Hill, 1994.

Brett Beemyn and Mickey Eliason. *Queer Studies: A Lesbian, Gay, Bisexual, & Transgender Anthology*. New York: New York University Press, 1996.

Bridges, Judith S. "Pink or Blue: Gender-Stereotypic Perceptions of Infants as Conveyed by Birth Congratulations Cards." *Psychology of Women Quarterly* 17 (1993):193-205.

David D. Gilmore. *Manhood in the Making: Cultural Concepts of Masculinity*. New Haven, CT: Yale UP, 1990.

Denmark, Florence, Vita Rabinowitz, and Jeri Sechzer. *Engendering Psychology*. Boston: Allyn and Bacon, 2000.

Erikson, Erik H. Identity, *Youth, and Crisis*. New York: Norton, 1968.

Fagot, Beverly I., and Mary D. Leinbach. "The Young Child's Gender Schema: Environmental Input, Internal Organization." *Child Development* 60 (1989):663-672.

Felicity Haynes and Tarquam KcKenna. *Unseen Genders: Beyond the Binaries*. New York: 2001.

Gordene Olga MacKenzie. *Transgender Nation*. Bowling Green, OH: Bowling Green University Popular Press, 1994.

Halpern, Diane F. *Sex Differences in Cognitive Abilities*, 3rd edition. Mahwah, NJ: Lawrence Erlbaum, 2000.

Holly Devor. *Gender Blending: Confronting the Limits of Duality*. Bloomington, IN: Indiana UP, 1989.

John Money & Anke A. Ehrhardt. *Man & Woman, Boy & Girl*. Baltimore, MD: Johns Hopkins UP, 1972.

John Money & Patricia Tucker. *Sexual Signatures: On Being a Man or a Woman*. Boston, MA: Little, Brown, 1975.

John Money. *Gendermaps: Social Constructionism, Femininism, and Sexosophical History*. New York: Continuum, 1995.

.... *Venuses Penuses:Sexology, Sexosophy, and Exegency Theory*. Buffalo, NY: Prometheus Books, 1986.

John Stoltenberg. *The End of Manhood: A Book for Men of Conscience*. New York: Dutton, 1993.

Kohlberg, Lawrence. "A Cognitive-Developmental Analysis of Children's Sex Role Concepts and Attitudes." In Eleanor E. Maccoby ed., *The Development of Sex Differences*. Stanford, CA: Stanford University Press, 1966.

Letty Cottin Pogrebin. Growing Up Free: Raising Your Child in the 80's. New York: McGraw-Hill, 1980.

Maccoby, Eleanor E. *The Two Sexes: Growing Up Apart, Coming Together*. Cambridge, MA: Harvard University Press, Belknap Press, 1998.

Marc Feigen Fasteau. *The Male Machine*. New York: McGraw-Hill, 1974.

Martin, Carol L., and Charles F. Halverson. "A Schematic Processing Model of Sex Typing and Stereotyping in Children." *Child Development* 52 (1981):1119-1134.

Myriam Miedzian. *Boys Will Be Boys: Breaking the Link Between Masculinity and Violence.* New York: Doubleday, 1991.

Pat Califia. *Public Sex: The Culture of Radical Sex.* San Francisco, CA: Cleis Press, 1994.

Pollack, William. *Real Boys.* New York: Henry Holt, 1998.

Rachel T. Hare-Mustin and Jeanne Marecek, eds. *Making a Difference: Psychology and the Construction of Gender.* New Haven, CT: Yale UP, 1990.

Rheingold, Harriet L., and Kaye V. Cook. "The Content of Boys' and Girls' Rooms as an Index of Parents' Behavior." *Child Development* 46 (1975):445-463.

Robert J. Stoller. Presentations of Gender. New Haven, CT: Yale UP, 1985.

Robert J. Stoller. *Sex and Gender: On the Development of Masculinity and Femininity.* New York: Science House, 1968.

Robert Moore and Douglas Gillette. *The King Within: Assessing the King in the Male Psyche.* New York: Morrow, 1992.

Sandra Lipsitz Bem. *The Lens of Gender: Transforming the Debate on Sexual Inequality.* New Haven, CT: Yale UP, 1993

Sandra Lipsitz Bem. *The Lens of Gender: Transforming the Debate on Sexual Inequality.* New Haven, CT: Yale UP, 1993.

Sherry B. Ortner and Harriet Whitehead, eds. *Sexual Meanings: The Cultural Construction of Gender and Sexuality.* Cambridge, UK: Cambridge UP, 1981.

Sherry B. Ortner and Harriet Whitehead. *Sexual Meanings: The Cultural Construction of Gender and Sexuality.* Cambridge, UK: Cambridge UP, 1981.

Stern, Marilyn, and Katherine H. Karraker. "Sex Stereotyping of Infants: A Review of Gender Labeling Studies." *Sex Roles* 20 (1989):501-522.

Susan Brownmiller. *Femininity.* New York: Linden Press/Simon & Schuster, 1984.

Susan Golombok & Robyn Fivush. *Gender Development.* Cambridge, UK: Cambridge UP, 1994.

Index

P

R

S

T

W